Gendered Violence at International Festivals

Gendered Violence at International Festivals is a groundbreaking collection that focusses on this highly important social issue for the first time. Including a diverse range of interdisciplinary studies on the issue, the book contests the widely held notion that festivals are temporal spaces free from structural sexism, inequalities or gender power dynamics. Rather, they are spaces where these concerns are enhanced and enacted more freely and where the experiential environment is used as an excuse or as an opportunity to victim blame and shame.

In this emerging and under-researched area, the chapters not only present original work in terms of topics but also in theoretical and methodological approaches. Each of the chapters are cross- or interdisciplinary, drawing on gender, sexualities, cultural and ethnicity studies. Studies from a range of highly regarded academics based around the world examine the subject by looking at examples from a wide range of destinations, including Spain, Argentina, Nigeria, Zimbabwe, Australia, Canada and the UK. This significant book progresses understanding and debates about gendered festival experiences and emphasises the symbolic and physical violence often associated with them.

This will be of great interest to undergraduate and postgraduate students and academics in the field of events studies. It will also be of use to practitioners or non-profit workers in the festival industries, including festival management organisations and planning committees.

Louise Platt is Senior Lecturer in Festival and Events Management at Manchester Metropolitan University.

Rebecca Finkel is Reader in Events Management at Queen Margaret University and Senior Fellow of Higher Education Academy.

Routledge Critical Event Studies Research Series
Edited by Rebecca Finkel
Queen Margaret University, UK
David McGillivray
University of the West of Scotland, UK

Gendered Violence at International Festivals
An Interdisciplinary Perspective
Edited by Louise Platt and Rebecca Finkel

For more information about this series, please visit: www.routledge.com/tourism/series/RCE

Gendered Violence at International Festivals
An Interdisciplinary Perspective

**Edited by Louise Platt
and Rebecca Finkel**

LONDON AND NEW YORK

First published 2020
by Routledge
2 Park Square, Milton Park, Abingdon, Oxon OX14 4RN

and by Routledge
52 Vanderbilt Avenue, New York, NY 10017

Routledge is an imprint of the Taylor & Francis Group, an informa business

© 2020 selection and editorial matter, Louise Platt and Rebecca Finkel; individual chapters, the contributors

The right of Louise Platt and Rebecca Finkel to be identified as the authors of the editorial material, and of the authors for their individual chapters, has been asserted in accordance with sections 77 and 78 of the Copyright, Designs and Patents Act 1988.

All rights reserved. No part of this book may be reprinted or reproduced or utilised in any form or by any electronic, mechanical, or other means, now known or hereafter invented, including photocopying and recording, or in any information storage or retrieval system, without permission in writing from the publishers.

Trademark notice: Product or corporate names may be trademarks or registered trademarks, and are used only for identification and explanation without intent to infringe.

British Library Cataloguing-in-Publication Data
A catalogue record for this book is available from the British Library

Library of Congress Cataloging-in-Publication Data
A catalog record for this book has been requested

ISBN: 978-0-367-36254-6 (hbk)
ISBN: 978-0-429-34489-3 (ebk)

Typeset in Times New Roman
by Apex CoVantage, LLC

Contents

About the contributors vii
Acknowledgements xii

1 **Gendered violence at international festivals: an interdisciplinary perspective** 1
   LOUISE PLATT AND REBECCA FINKEL

2 **Analysis of the response of the feminist movement and institutional feminism to gender violence in local festivals in northern Spain: the case of the Basque Country** 9
   MARÍA SILVESTRE, RAQUEL ROYO AND ESTIBALIZ LINARES

3 ***Fiestas*, public space and rape culture. A study of the Wolf Pack case** 24
   ANNA MORERO BELTRÁN AND CLARA CAMPS CALVET

4 **Sexual violence in Nigerian MusiCultural festivals: narratives of the unheard victims** 39
   RICHARD A. ABORISADE

5 **Between patriarchy and capitalism: the gendered violence of Intwasa International Arts Festival, Bulawayo, Zimbabwe** 54
   KHANYILE MLOTSHWA

6 **Gender, transgression and sexual violence at Australian music festivals** 69
   BIANCA FILEBORN, PHILLIP WADDS AND STEPHEN TOMSEN

7 **Conceptualising safety and crime at UK music festivals: a gendered analysis** 86
   HANNAH BOWS, HANNAH KING AND FIONA MEASHAM

8   Gender-based violence amongst music festival employees   104
    CASSANDRA JONES

9   Structural disputes: an analysis of infrastructural inequalities
    in the case study of the Women of the World Festival in Hull
    (UK) City of Culture 2017   119
    BARBARA GRABHER

10 The green wave! Resistance in festivals as part of the women
    empowerment process in Argentina   133
    VALERIA VEGH WEIS AND LUCIA MONTENEGRO

11 The polite cowboy or the wild, wild west: strategic approaches
    to reducing gender-based rodeo violence through grassroots
    civic mobilisation   147
    MYLYNN FELT AND MARIA BAKARDJIEVA

    Index   163

About the contributors

Richard A. Aborisade, PhD, is Senior Lecturer of the Department of Sociology, Faculty of Social Sciences, Olabisi Onabanjo University, Ago-Iwoye, Ogun State. He received his doctorate from the University of Ibadan, Nigeria. He also holds an MBA in information technology from Coventry University, UK. He has published in both local and international journals in the areas of security management, criminal justice, criminology, victimology and penology. His most recent work includes *The Essentials of Sociology* (co-edited) and *Crime and Delinquency: A Sociological Introduction*, both published by Ibadan University Press.

Maria Bakardjieva, PhD, is Professor in the Department of Communication and Culture, University of Calgary. She is the author of *Internet Society: The Internet in Everyday Life* (2005, Sage) and co-editor of *Socialbots and Their Friends* (2017, Routledge) and *How Canadians Communicate* (2004 and 2007, University of Calgary Press). She held the position of Editor in Chief of the *Journal of Computer-Mediated Communication* from 2011 to 2013. Her work on the topics of Internet use in everyday life, online community, e-learning and research ethics has been published in journals such as *Media, Culture and Society*; *New Media and Society*; *The Information Society*; *Philosophy and Technology*; *Ethics and Information Technology* and *Sage Benchmarks in Communication*, amongst others.

Anna Morero Beltrán holds a PhD in sociology and a master's degree in women, gender and citizenship. She is based at the University of Barcelona where she is Associate Lecturer in the Department of Sociology. She is a member of the Inter-University Research Group COPOLIS: "Welfare, Community and Social Control" and the Inter-University Women and Gender Studies Institute. Her research interests are related to gender-based violence and the social study of reproductive and genetic technology; she has carried out several research projects and published various scientific and informative articles in these fields.

Hannah Bows, PhD, is Assistant Professor of criminal law at Durham Law School, Durham University. She is Deputy Director of the Centre for Research into Violence and Abuse. Her work broadly focusses on violence against women. Over the last seven years, most of this work has been concerned with violence against women aged 60 and over. She is currently the principle investigator on a British Academy–funded study examining sexual violence at UK music festivals. She has also been awarded a three-year British Academy fellowship to examine criminal justice outcomes and responses to crimes involving older victims. Outside of the university, she is Chair of the British Society of Criminology Victims Network and currently Chair of Age UK Teesside.

Clara Camps Calvet is Associate Lecturer of sociology at the University of Barcelona and the University of Girona. Her main lines of research are gender, social movements and social change, as well as punitive control and political repression. She has published numerous articles on these issues. She is a member of the Inter-University Research Group COPOLIS: "Welfare, Community and Social Control" at the Department of Sociology at the University of Barcelona. She is involved in different social movements in Barcelona.

Mylynn Felt is a PhD candidate in the graduate programme in communication at the University of Calgary. She is a Vanier Scholar and the author of "Social Media and the Social Sciences: How Researchers Employ Big Data Analytics" in the journal *Big Data and Society*. She has also published articles with *Information, Communication & Society*; *The Canadian Journal of Communication*; and elsewhere. Her dissertation research focusses on grassroots civic engagement utilising social media to impact gender-based violence. Her past research examined the news framing of cyberbullying as a social problem.

Bianca Fileborn, PhD, is Lecturer in criminology at the School of Social and Political Sciences, University of Melbourne. Her research examines the intersections of space/place, identity, culture and sexual violence and justice responses to sexual violence. She was awarded an Australian Research Council Discovery Early Career Researcher Award in 2019 to examine justice responses to street harassment. She is the author of *Reclaiming the Night-Time Economy: Unwanted Sexual Attention in Pubs and Clubs* (Palgrave Macmillan) and co-editor of *#MeToo and the Politics of Social Change* (Palgrave Macmillan).

Rebecca Finkel, PhD, is an urban cultural geographer and Reader in Events Management at Queen Margaret University, Edinburgh, and Senior Fellow at the Higher Education Academy. The main focus of her research frames critical event studies within conceptualisations of social justice, gender in/equality and cultural identity. Her primary research interests include resistance to globalisation processes through cultural events, doing gender at festivals and mapping human rights and international sporting events. She is committed to accessibility, inclusion and diversity in education through her teaching practices and research focus.

Barbara Grabher is Research Assistant at the Culture, Place and Policy Institute and PhD candidate in the context of the Horizon 2020 Marie Skłodowska-Curie

About the contributors ix

Action Framework GRACE project (Gender and Cultures of Equality in Europe) at the University of Hull, UK, and University of Oviedo, Spain. She holds a BA in cultural and social anthropology from the University of Vienna, Austria, and an MA in gender studies from Utrecht University, Netherlands, and University of Granada, Spain.

Cassandra Jones, PhD, is an early career researcher whose interest focus on gender-based violence, men and masculinities and mixed-method evaluations of gender-based violence intervention and prevention programmes. She has over ten years of experience in the US and UK using both quantitative and qualitative methods to provide unique insight into the experiences and use of gender-based violence amongst marginalised groups. Her PhD at the University of Bristol highlighted the intersectional complexities in the relationship dynamics of men who identified as victims of domestic violence and abuse and the paradoxical language they used to describe their experiences.

Hannah King, PhD, is Assistant Professor in the Department of Sociology at Durham University. She draws upon her previous policy and youth work practice expertise within her teaching and research. Her research interests are in domestic and sexual violence and young people's experiences of marginalisation and youth policy. In recent years, she has been engaged in festivals research, including sexual violence at festivals. She is committed to working with innovative and participatory methodologies. She also co-directs the Inside-Out Prison Exchange Programme at Durham.

Estibaliz Linares holds a PhD in the programme in Human Rights: Ethics, Social, and Politics Challenges, with "The Macho-Driven Digital Iceberg" project, for which she obtained the Cum Laude qualification. She has a social work degree and intervention against violence against women postgraduate certification from University of Deusto. She is Deusto Social Values Researcher and Intervention in Violence against Women Postgraduate from the University of Deusto. Now, her main research lines are focussed on gender, adolescence and virtual life. In addition, she collaborates as a trainer of prevention of violence amongst teenagers in Sortzen S.l consultancy and the Ctrla programme.

Fiona Measham, PhD, is Professor and Chair in criminology at the University of Liverpool. She has conducted research across three decades exploring changing trends in drugs and their regulation and policing and festivals, nightlife and the sociocultural context to drugs. Her service on scientific advisory committees includes the Advisory Council on the Misuse of Drugs 2008–2018, the Ministerial Review of New Psychoactive Substances and Labour, Liberal Democrat and Conservative Party drug policy expert panels. She is a permanent member of DrugScience. She is co-director of the Loop and the Loop Australia, non-governmental organisations providing harm reduction and drug-checking services in festival and nightlife settings.

Khanyile Mlotshwa is completing his PhD (media and cultural studies) at the University of KwaZulu-Natal, Pietermaritzburg campus, South Africa. In his research, he experiments with transdisciplinary approaches in urban, migration, border, media and cultural studies to interrogate the discursive constructions of black African subjectivity in post-apartheid South Africa. He has published work in peer-reviewed journals that include *Agenda*, *Cross-Cultural Human Rights Review* and the *Westminster Papers in Communication and Culture*. He has chapters forthcoming in edited book collections. He has presented at conferences in South Africa, Mozambique, Senegal, Ghana, Germany and the Netherlands in the past two years.

Lucía Montenegro, BA in law, LL.M candidate, is an Argentinian professor of criminal law and criminal procedure at Buenos Aires University. She works at the Public Defense Office in Argentina. In addition, she is a member of the Law Female Professors' Network at Buenos Aires University. She co-coordinated the publication of their first book, *Abortion: The Green Wave from the Law's Perspective*. Her research themes include constitutional rights and feminist theory applied to criminal law and criminal procedure, particularly access to justice, criminal response to gender-related offences, abortion and the right to self-defence in cases of gender violence.

Louise Platt, PhD, is Interdisciplinary Researcher and Senior Lecturer in Festival Management at Manchester Metropolitan University. She is a fellow at the Institute of Place Management and a member of the Executive Committee of the Leisure Studies Association. Her research focus is on placemaking and festivity with a particular concentration on processional forms and experiences of festivals and leisure spaces. Her work predominantly draws on cultural geography, dance/performance theory and poststructural philosophy to elucidate a more fluid understanding of place and community using festivity as a lens. She teaches festival studies at both the undergraduate and postgraduate levels and supports PhD students on festival-related topics.

Raquel Royo holds a PhD in sociology and social work degree. She is Director of Intervention against Violence against Women postgraduate at University of Deusto. She takes part in the PhD programme of Social Sciences and Humanities Faculty in which she teaches Discrimination and Gender Violence and Gender Sociology. She is a member of Deusto Social Values research team, which analyses the European Values Questionnaire. She is Principal Researcher in diverse researchers founded by the Emakunde-Women Basque Institute and Biscay Council. Her principal lines of specialisation are motherhood, fatherhood and co-responsibility, as well as value and gender.

María Silvestre holds a PhD in political sciences and sociology. She is Main researcher of the Deusto Social Values research team, which represents Spain in the European Values Studies. She has been Dean of Political Sciences and Sociology at the University of Deusto (2004–2009) and Director of Intervention against Violence against Women postgraduate (2003–2009). She has been Director of the Emakunde-Women Basque Institute (2009–2012). She has led various research projects in competitive and concerted calls and assumed different management responsibilities, such as the Basque Sociology Association presidency. Her principal specialisation areas are political-social values and gender perspective in the social sciences, which are areas in which she has publications.

Stephen Tomsen is Professor of Criminology at the University of Western Sydney and has held visiting professorships at universities in the UK, US and the Netherlands Ministry of Justice. He was a pioneering figure in the international development of three key areas in contemporary criminology: nightlife ethnographies, queer criminology and crime and masculinity studies, which developed as new academic fields in the 1990s and early 2000s. He is well-known for his research and publications on violence, homicide, hate crime, victims, gender and sexuality, drinking and drug use and debates regarding urban public order.

Valeria Vegh Weis, LL.M, PhD, is an Argentinean/German professor of criminology and transitional justice at Buenos Aires University and National Quilmes University. She is currently Associate Researcher at the Max Planck Institute for European Legal History and an Alexander von Humboldt Post-Doctoral Researcher at Freie Universität Berlin. She holds a permanent position at the Argentinean Public Defense Office. She has 15 years of experience working in criminal courts and international organisations. Her book *Marxism and Criminology: A History of Criminal Selectivity* received awards by the American Library Association and the Academy of Criminal Justice Sciences. She is a member of the Law Female Professors' Network at Buenos Aires University.

Phillip Wadds, PhD, is Senior Lecturer in criminology at University of New South Wales (UNSW) Sydney. His research is situated at the intersection of four interrelated themes: (1) policing, (2) nightlife and related leisure, (3) alcohol and other drugs (AODs) and (4) violence. He has spent the last decade undertaking field-based research examining various features of nightlife in Sydney with an enduring focus on its policing and regulation. More recently, he has conducted world-first research with Bianca Fileborn and Stephen Tomsen on safety and sexual violence at Australian music festivals.

Acknowledgements

This book is powerful and meaningful due to the dedicated research of the chapter authors who have undertaken important and difficult work to ensure that the implications of gendered experiences at festivals are more fully understood. Thank you! And we look forward to future collaborations that catalyse positive social change in the critical events field.

Many thanks to the team at Routledge, Emma Travis and Lydia Kessell, for their support.

We dedicate this book to all the unheard voices who have experienced gendered violence at international festivals. We're now listening. We now demand action.

1 Gendered violence at international festivals

An interdisciplinary perspective

Louise Platt and Rebecca Finkel

Introduction

The idea for this book first started out as a chapter for Lamond and Moss's (forthcoming) volume exploring liminality in critical event studies. As we developed our chapter, we realised the topic of the liminal/liminoid was so much more problematic than we initially anticipated once the focus shifted to non-dominant, non-privileged groups and cultures. How could ingrained inequalities, un/conscious bias and motivations for symbolic, as well as physical, violence just float away when people enter festival spaces? And then like an overcoat, would these prejudices be put on again as they walked out of these so-called liminal environments? Although festivals are often marketed as 'transformational', this mainly centres on the experiential design and programming content rather than the interactions between participants and attendees. It is too big of an ask for festivals alone to dismantle societal power structures; in fact, more often than not, they reinforce them. As Finkel et al. (2018, 1) stated, "Special events are microcosms of society. Because they are temporary and usually bound by geographic space, they can be considered reflections of or responses to societal norms at the time they take place".

We thought other scholars must have realised this too, and thus we wanted to find out what interdisciplinary work was being conducted on a global scale. Our book chapter grew into an edited book; it is the first in the Routledge Critical Event Studies Research Series. As co-editors of the series, Finkel and McGillivray state,

> In adopting a critical stance, we take our starting point the idea that events intervene in social structures and, in so doing, they expose the diverse contested discourses and frames of reference being articulated by the actors and institutions involved.

Given this, we specifically wanted to focus on gender – not necessarily broader equality and diversity narratives, which can often entangle many disparate communities together without going into depth about distinctive key issues or how these may interrelate. Certainly, we do recognise that all useful discussions surrounding gender are intersectional (Crenshaw, 1989), and, as such, the ways in which age, class, nationality, ethnicity and so forth intersect with gender are of particular relevance to advancing research in this area. Thus we wanted to concentrate on how intersectional gendered relations play out in festival spaces and the sometimes violent outcomes as a result of these interactions. Although many current conversations of gendered violence are viewed as a binary with most discussions highlighting men's violence towards women, we would like to state that we accept gender as a spectrum and an identity culturally developed and performed, aligned with Butler's (1999) theorisations. We also acknowledge that much more investigation is needed regarding those with trans and non-binary gender identities in experiential environments, and there is significant scope there for future research.

Although more work is being published at the intersections of gender studies and critical event studies (e.g. Coyle & Platt, 2018; Crichton & Finkel, 2019; Dashper, 2013, 2018; Finkel & Dashper, 2020; Platt & Finkel, 2018; Pavlidis, 2012; Walters, 2018), it is still underexplored, emerging and considered rather niche, often sidelined to its own segregated conference streams and special issues instead of embraced as an integral theme threaded through mainstream scholarship (Dashper & Finkel, 2020). Despite a growing body of work regarding human trafficking and sex work around mega sporting events (e.g. Finkel & Finkel, 2015; Matheson & Finkel, 2013; De Lisio et al., 2019), it is still rare for the unfavourable, less salubrious actions of human behaviour to be researched within festival frameworks, with very few publications concentrating on sexual assault, rape and other incidences of violence. Therefore, we seek to make a contribution by amplifying more realistic festival narratives to inform conceptualisations and practices in an effort to address pervasive gendered injustice issues.

Thus the emphasis for this volume centres on patriarchy, performativity and praxis in an effort to contest the widely held notion that festivals are temporal spaces free from structural sexism, inequalities or gender power dynamics. Instead, we argue that they are spaces where these are enhanced and enacted more freely, using the experiential environment as an excuse or as an opportunity to victim blame and shame. Whilst it has been argued that the liminal/liminoid space, due to the so-argued temporary suspension of societal norms, allows women's bodies and voices to be asserted in a way that is not usually deemed acceptable in everyday public spaces (e.g. Riches, 2011), there is often neglect in understanding how hegemonic social and cultural structures and controls still govern these spaces. This book illustrates this argument with discussions around the increase in reported sexual assaults at international festivals to argue that a persistence to characterise festivals spaces as uncomplicated, value-free, utopic liminal/liminoid is highly problematic.

The problem with the liminal-norm: festivals and gendered violence

We argue that liminality as conceived by van Gennep (1960) and, subsequently, Turner (1969, 1979, 1982) within the festival literature has been under-theorised and, as a result, has limited event scholars' abilities to be critical of festival spaces. Whilst it has been identified that the liminal space of the festival can be dangerous and risky (e.g. Jaimangal-Jones et al., 2010), there is little work that questions issues of gender power relations within festival frameworks. Without wanting to spill further ink on explaining what liminality is, we want to point out briefly some key aspects of Turner's work that have been acknowledged by festival scholars which are integral to our analysis. In festival studies, the work of Falassi (1987) is often cited in relation to his take on the ritual process drawn from van Gennep, whereby the festival space is seen as "a time out of time". Within the festival space, in this liminoidal period, *communitas* occurs – i.e. festival goers come together with the same purpose and become a "communion of equal individuals" (Turner, 1969, 96). Turner's characterisation of structure and anti-structure has been appealing to those studying festival spaces, particularly with an emphasis on transgression, freedom and creative expression. Yet there is an assumption that power is dispersed or even absent under such 'communitas'. Boissevain (2016), however, notes that festivals evolve within the changing nature of society and, therefore, are not immune to power relations external to the events themselves. Within festival studies specifically, there is an emerging critique; most relevant here is Pielichaty's (2015) work in relation to the negotiation of gender in festival experiences. In her analysis, she found that despite the so-called liminoid spaces of festivals, gender roles were normative and reinforced. This volume expands on her important work to examine what the consequences of this might be.

The lack of critique in festival studies of the limen and communitas, we contend, has occurred as festival scholars have fallen under the spell of the 'liminal-norm' (McKenzie, 2004), a theory from performance studies that suggests that the subversive potential of performance has become the status quo, with a privileging of liminality leading to conceptualisations of the term becoming chronic and normative. In a critique of the concept of liminality in performance studies, Crosby (2009, 5) suggests that the concept of the liminal, "produces a blindness for, or at least impatience with, institutions and cultural performances associated with reproduction and structure rather than resistance or subversion". Therefore, there is the accusation that by employing the concept as an analytical tool, we can be in danger of being too ready to seek subversion and not address the structures within which hegemonic society functions. McKenzie (2004) considers the symbiotic relationship between liminality and performance and suggests that liminality as a theoretical conceptualisation is essential to understand performance as a transgressive force. Indeed, McKenzie (2004) further points out that Turner even accepted that liminal rituals had normalising functions on society, and the outcome of ritual is often 'reabsorption' into society.

> Public liminality can never be tranquilly regarded as a safety valve, mere catharsis, 'letting off steam'. Rather, it is communitas weighing structure, sometimes finding it wanting, and proposing in however extravagant a form new paradigms and models which invert or subvert the old.
>
> (Turner, 1979, 474)

McKenzie (2004) goes on to argue that liminality was 're-cited' and 'de-contextualised' throughout the 1960s and '70s in performance theory with the rarer moments of rupture in ritual being centralised. Whilst problematic, what McKenzie (2004) is suggesting is that the notion of liminality cannot be abandoned altogether; however, there is a need to acknowledge its 'own alterity' in order to unravel the reification of its theorisation.

Thus the ways in which festival spaces have been perpetually conceptualised in this regard motivated us, as editors of this volume, to call for international perspectives on how gendered violence is enacted across different festival environs.

Chapter outline

Recently, there have been more and more media reports surrounding increased sexual assaults at cultural and, specifically, music festivals. However, there has been a paucity of sustained scholarly work in this area. This collection attempts to highlight international, interdisciplinary research in an effort to progress thinking about gendered festival experiences and emphasise the symbolic and physical violence often associated with them. The chapters draw from a range of sociology, geography, criminology, law, cultural studies and other social sciences and humanities literatures and methodologies, along with gender studies and critical event studies. This book is in no way conclusive or all encompassing, as there is scope for much more research into this area; however, it is, in some ways, a spark for an ongoing conversation about gendered violence (both epistemic and physical) at festivals and the implications this has for the ways we celebrate – and live – together. The vision for this book, then, is to feature contributions from critical events and interdisciplinary scholars specifically focussed on all kinds of gendered violence at festivals from around the world. Each chapter interweaves cross-disciplinary theories and international policies and includes practical case study example/s to illustrate key points. Unfortunately, almost all continents are represented in these chapters, which illustrates what a pervasive and universal issue this is, and provides further evidence of the need for such gendered inequalities to be more competently addressed – inside and outside – of the festival setting.

The book begins by exploring the manifold dimensions of the infamous *La Manada* or 'Wolf Pack' gang rape case in Spain. Silvestre, Royo and Linares analyse the responses of feminist activist movements to this horrific sexual assault. The chapter presents political and social feminist interventions against gendered violence and sexual aggressions that occur in the popular festivals of the Basque Country in northern Spain. Festival lived experiences are contrasted with institutionalised measures to demonstrate the increasing internalisation of empowering feminist discourses in younger generations of women. Following this, Anna Morero Beltrán and Clara Camps Calvet also focus their chapter on the 'Wolf Pack' gang rape case by demonstrating how the event came to mark a turning point in the country's responses to sexual violence in general and, more specifically, to sexual attacks perpetrated in public space during celebratory events. They argue that a 'patriarchal logic' dominates the uses of public space, and *fiestas* function not merely as leisure activities but mainly as practices that represent the current social order, legitimise violence against women, reinforce sexist hierarchical values and reproduce rape culture.

The next two chapters focus attention on African case study examples, where similar themes emerge relating to how broader societal inequalities are magnified in cultural festival spaces and how this results in a continuous cycle of gendered violence. Aborisade's research explores sexual aggression against women by men in cultural festival spaces in southern Nigeria. The extent of gendered harassment and attacks are indicative of broader social heteronormative ordering in Nigerian society and illustrate the safety issues women must deal with on a regular basis. Thereafter, Mlotshwa examines the gendered epistemic violence and colonialism of the content, programming schedules and location of shows at an arts festival in Zimbabwe and how the festival, perhaps unintentionally, contributes to the symbolic degradation of women and persistence of colonialism of gender.

Next, we turn the spotlight on music festivals. Fileborn, Wadds and Tomsen investigate gendered practices which lead to sexual violence in an Australian context. They explore the potential of 'assemblage thinking' in understanding how individual incidents of sexual violence unfold in music festival settings. Also, from a criminologist perspective, Bows, King and Measham analyse gendered safety at UK music festivals. Along with studying the occurrence and nature of crime, they also examine perceptions, concerns and experiences of different types of gendered harassment and violence at UK music festivals. Most of these chapters are examined from consumers' perspectives, but Jones flips this by researching producers' perspectives with an analysis of workplace gender-based violence at UK music festivals. It is important to understand that it is not just audiences who are affected by gendered violence at festivals; the experiences of festival employees are often neglected and require further attention and intervention to create safer and more respectful working environments for festival staff by untangling the gendered power dynamics of music festival organisational structures. Additionally, the experiences of performers is not an issue covered in-depth in this book, but it should in no way be discounted, and further research from performers' perspectives would be very welcome and necessary for providing more evidence, experiences, lessons learned and best practices to shape our understanding of gendered violence at festivals and, thus, provide better practical solutions.

Grabher explores the micro-, meso- and macro-infrastructural inequalities of a women's festival as part of a wider city of culture series of events in Hull, UK. With attention to structural discrimination in so-called equality-themed events, she probes the fragility of the notion of equality. Vegh Weis and Montenegro, then, interrogate the social, economic and cultural contexts of festivals in Argentina through the actions of feminist activists who are drawing attention to the sexual violence and blatant sexism of the festival scenes there. Such resistance is not only catalysing changes in Argentine festival landscapes but also calling into question societally accepted notions of inequitable gendered behaviour and attitudes.

The book concludes with a case study example of a heritage sporting festival, demonstrating that it is not just cultural and music festivals where gendered violence is an issue; rather, it is an issue in all kinds of festivals when they are reflections of societal values and biases. Felt and Bakardjieva base their research on the Calgary Stampede in Canada, where they found that an often-uncritical view of a celebration of Western heritage has led to a culture normalising of sexual harassment and other forms of gender-based violence during the festivities. A grassroots social movement has responded to this in an effort to raise awareness and make the festival safer. The chapter explores how grassroots civic mobilisations can influence festival culture and practices pertaining to gender politics and put gender-based violence on the agenda for the festival's organisers.

Future directions for research and practice

As is evident throughout this volume, there is momentum building by feminist activist movements to take action against symbolic and physical violence in festival spaces throughout the world. Internationally, harassment and sexual attacks are no longer considered something women "just have to put up with" and "a risk they have to take" to enjoy celebratory spaces. It is not acceptable, and although more encompassing change to eradicate structural inequalities in global cultures will be slower to happen, it is within our power to change how gendered interactions occur at festivals. Festivals are not naturally misogynistic entities. And although they are mainly currently *reflections* of society, festivals do have the potential to be constructed as *responses* to society to make statements about social needs and social anxieties (Warren, 1993). As Finkel and Dashper (2020) argued, "Events can provide forums to advance alternative social arrangements and to work as a force for society's transformation". Although they cannot be a panacea for societal woes, festivals can take some steps in progressing positive social change. Therefore, what we can conclude from the chapters in this book, which represent a snapshot of what is happening again and again in festival after festival, is that more must be done by events managers, local authorities and police, community groups and sociocultural institutions to change the physical and cultural infrastructure enabling gendered violence in experiential landscapes. There can no longer be a prioritisation of the economic over the social when making decisions about festival design, programming, location, security, alcohol provision and so forth without consideration of the implications for gendered violence. Festivals must be co-created with approaches which take into account dominant power frameworks – until the unjust hierarchies of mainstream society are dismantled. Audiences, workers, performers – all stakeholders involved in festival industries – have to be afforded the same opportunities to enjoy safe, nonviolent and, then quite possibly, transformative festival experiences.

Therefore, much more research into gender interactions in festival spaces is needed, including, as mentioned earlier, trans and non-binary experiences, as well as diverse viewpoints from various festival stakeholders. Joined up thinking amongst interdisciplinary researchers, including shared approaches and cross-subject collaborations, have the potential to provide multi-perspective narratives, which are helpful in developing in-depth critical analyses of the complex dynamics in festival settings. As Finkel and Dashper (2020) argued, "Critical event research has the potential to inform current theoretical developments and wider sector practices, and, ultimately, change the dominant heteronormative patriarchal paradigm of the experiential landscape". We hope that this book instigates wider academic, policy, management and organisational debates about these issues in this context.

What we have assembled here are narratives of dis/empowerment with a view to enacting positive social change. We consider this to be a dawning of evolving conversation amongst scholars and practitioners, where festivals are not continuously framed in unrealistic, outmoded, patriarchal conceptualisations of communitas and decisions can be made to construct these celebratory spaces in ways that acknowledge and amend existing gender inequalities. As Audre Lorde (1984) so aptly said, "And then our speaking out will permit other women to speak, until laws are changed and lives are saved and the world is altered forever".

References

Boissevain, J. (2016). The dynamic festival: Ritual, regulation and play in changing times. *Ethnos*, *81*(4), 617–630.

Butler, J. (1999). *Gender trouble: Feminism and the subversion of identity* (2nd ed.). Abingdon: Routledge.

Coyle, T., & Platt, L. (2018). Feminist politics in the festival space. In J. Mair (Ed.), *Routledge handbook of festivals*. Abingdon: Routledge.

Crenshaw, K. (1989). Demarginalizing the intersection of race and sex: A Black feminist critique of antidiscrimination doctrine, feminist theory and antiracist politics. *University of Chicago Legal Forum*, *140*, 139–168.

Crichton, A., & Finkel, R. (2019). Barriers to access: Investigation of plus-size women consumer experiences at fashion events. In T. Walters & A. Jepson (Eds.), *Marginalization and events*. Abingdon: Routledge.

Crosby, J. (2009). Liminality and the sacred: Discipline building and speaking with the other. *Liminalities: A Journal of Performance Studies*, *5*(1), 1–19.

Dashper, K. (2013). The "right" person for the job: The aesthetics of labour within the events industry. *Event Management*, *17*(2), 135–144.

Dashper, K. (2018). Confident, focussed and connected: The importance of mentoring for women's career development in the events industry. *Journal of Policy Research in Tourism, Leisure and Events*, *10*(2), 134–150.

De Lisio, A., Hubbard, P., & Silk, M. (2019). Economies of (alleged) deviance: Sex work and the sport mega-event. *Sexuality Research and Social Policy*, *16*(2), 179–189.

Falassi, A. (1987). *Time out of time: Essays on the festival*. Albuquerque, NM: University of New Mexico Press.

Finkel, R., & Dashper, K. (2020). Accessibility, diversity and inclusion in events. In S. Page & J. Connell (Eds.), *Routledge handbook of events*. Abingdon: Routledge.

Finkel, R., & Finkel, M. (2015). The "dirty downside" of global sporting events: Focus on human trafficking for sexual exploitation. *Public Health*, *129*(1), 17–22.

Finkel, R., Sharp, B., & Sweeney, M. (Eds.). (2018). *Accessibility, inclusion, and diversity in critical event studies*. Oxford: Routledge.

Jaimangal-Jones, D., Pritchard, A., & Morgan, N. (2010). Going the distance: Locating journey, liminality and rites of passage in dance music experiences. *Leisure Studies*, *29*(3), 253–268.

Lamond, I., & Moss, J. (Eds.). (forthcoming). *Exploring liminality in critical event studies: Boundaries, borders and contestation in the study and analysis of events*. London: Palgrave Macmillan.

Lorde, A. (1984). *Sister outsiders: Essays and speeches*. London: Penguin Books.

Matheson, C., & Finkel, R. (2013). Sex trafficking and Vancouver Olympic Games: Perceptions and preventative measures. *Tourism Management*, *36*, 613–628.

McKenzie, J. (2004). The liminal-norm. In H. Bial (Ed.), *The performance studies reader* (pp. 26–31). London: Routledge.

Pavlidis, A. (2012). From Riot Grrrls to roller derby? Exploring the relations between gender, music and sport. *Leisure Studies*, *31*(2), 165–176.

Pielichaty, H. (2015). Festival space: Gender, liminality and the carnivalesque. *Journal of Event and Festival Management*, *6*(3), 235–250.

Platt, L., & Finkel, R. (2018). Editorial: Special issue in equality & diversity in professional planned events industry. *Journal of Policy Research in Tourism, Leisure and Events*, *10*(2), 113–116.

Riches, G. (2011). Embracing the chaos: Mosh pits, extreme metal music and liminality. *Journal for Cultural Research*, *15*(3), 315–332.

Turner, V. W. (1969). *The ritual process: Structure and anti-structure*. Abingdon: Routledge.

Turner, V. W. (1979). Frame, flow and reflection: Ritual and drama as public liminality. *Japanese Journal of Religious Studies*, 465–499.

Turner, V. W. (1982). *From ritual to theatre: The human seriousness of play*. New York: PAJ Publications.

Van Gennep, A. (1960). *The rites of passage*. Chicago: University of Chicago Press.

Walters, T. (2018). Gender equality in academic tourism, hospitality, leisure and events conferences. *Journal of Policy Research in Tourism, Leisure and Events*, *10*(1), 17–32.

Warren, S. (1993). "This heaven gives me migraines": The problems and promise of landscapes of leisure. In J. Duncan & D. Ley (Eds.), *Place/culture/representation* (pp. 173–186). Abingdon: Routledge.

2 Analysis of the response of the feminist movement and institutional feminism to gender violence in local festivals in northern Spain
The case of the Basque Country

María Silvestre, Raquel Royo and Estibaliz Linares

Introduction

Festivals in the towns, neighbourhoods and cities of the Basque Country (northern Spain) are collective activities that have a markedly popular character – life is lived in the streets by day and night – and a notable capacity for channelling citizen participation. Besides institutional organisation, groups of different types play an active role in planning and developing numerous activities open to social participation for children, cultural and ludic events; the sale of food and drink; and competitions, without which it would be difficult to understand the idiosyncrasy of these festivals. This dynamism and collective construction configure these festivals as relevant sociocultural practices in the social life of the territory.

Popular festivals are also relational spaces for reproducing – and subverting – gender systems. According to a survey realised with 1,129 women on different forms of sexual violence in the context of nocturnal leisure in the Spanish state (Observatorio Noctámbul@s 2018), during the previous year, the great majority of those interviewed (97%) experienced reiterated forms of verbal violence with respect to their bodies or sexuality, non-consensual sexual touching (80%) and persistent male sexual attention (86%) while taking part in festivals. Additionally, 44% of these women said they were cornered by several men, 17% reported that they were raped without the use of force, while 5% were raped forcefully. The festival (whether popular or not) is at the top of the list of scenarios where these rapes occur – 78% of the interviewees endorse the idea that the festival is a "space totally exposed to these situations"; second in the list are the moments immediately following the festival (63%).

In this context, this chapter presents the political and social feminist interventions against gender violence and sexual aggressions that occur in the popular festivals of the Basque Country in northern Spain. A clear example is the media case of 'La Manada', which happened at the San Fermín Festival in 2016. This refers to the gang rape of a young woman by five men (known as 'La Manada' – the Wolf Pack). In the first instance, two courts in Navarre considered the events to be a crime of continuous sexual abuse (not aggression), as there had been no violence or intimidation. This ruling unleashed a wave of social indignation and feminist mobilisations in Navarre and the whole country under the slogan, "It's not abuse; it's rape". Subsequently, the Supreme Court considered that there had been sexual aggression and increased the sentence to 15 years. This sexual assault was met with a swift social and political rejection, but this response should not be considered spontaneous. In fact, it was related to several institutional actions and, above all, to the pressure exerted previously by the feminist movement. Consequently, the actions of the feminist movement and the performance of institutional feminism are set out next (especially that of Berdinsarea, a network of municipalities for equality between women and men founded in 2006).

The chapter concludes by presenting the evaluations of young women from the Basque Country of their experiences of participating in these festivities. This provides an element for contrasting the institutionalised measures adopted and serves as a way of gauging the internalisation of empowering feminist discourse in the younger generations of women.

Bodies, violence and gender: the social construction of sexual violence against women

Violence against women, including sexual access to their bodies without their consent, "is a constant fact in patriarchal society and culture" (Lagarde 2011, 278; Segato 2003, 24). Such violence forms part of a system of domination (Brownmiller 1975; Connell 1995) that has material and symbolic foundations and places women in a subordinate position in different orders of life – sexual, economic, labour, political, religious, reproductive, etc. From the feminist perspective, in spite of the differences of focus, the unequal distribution of power between women and men is considered the ultimate – not necessary the immediate or most visible – cause of this violence, insofar as it establishes and legitimises male control (symbolic and functional) over women (Sortzen 2011). That is why, as Lagarde (2012, 241) argued, "throughout most of history, women's bodies have been a space of domination, violence, and alienation. Indeed, women's bodies have been – and for many still are – occupied territories".

The socialisation of gender results in an overvaluation of what has been socially elaborated as masculine and in a logic of male imposition, with the cultural counterpart of female inferiority (Osborne 2009). Women have been historically constructed from a perspective of imminence and alterity (de Beauvoir 2000, 50, 63) as "rapeable beings" (Lagarde 2011, 295), as "bodies-of-others" and "for-others" (Basaglia 1983; Lagarde 2012, 244), who are not owed any reciprocity (Osborne 2009). From this perspective, sexual violence "is a political fact that synthesizes in an act the oppression and reification of women and the extreme realization of the patriarchal male condition". Rape is "the supreme fact of patriarchal culture"; its nucleus is power and the erotic appropriation of women, and it contributes to the cultural reproduction of gender as a whole and of the patriarchal male-female relations (Lagarde 2011, 279, 281–284).

The deep-rootedness of this type of violence is related to the fact that it represents one extreme of a continuum 'of normality' that exacerbates the traditional models that hold sway in our culture of masculinity-aggressiveness activity and femininity-sweetness passivity (Osborne 2009, 64, 80). Or stated differently, it is the subjugation and conquest of the sexual object (Giddens 2000), the 'mentality of conquest' prescribing that men should pursue women – and boast about this as a 'pact' amongst men – and that women resist (Osborne 2009, 65).

Brownmiller (1975, 5) considers rape to be an exercise of power that perpetuates the domination of women by keeping "all women in a state of fear". It is not the isolated act of a sick individual but a form of "patriarchal control", a "curfew for the whole female collective who find that their mobility is reduced: there are places and times into which no decent women venture" (Puleo 2005, 44–45). Rape attacks the individual and collective freedom of women, since not only the act of rape but also the fear of being raped constitutes a palpable representation of the inferiority and objecthood imposed on women and their lack of autonomy, which acts as a mechanism of dependence and subjection to male control (Osborne 2009). In the words of Ahmed (2018, 44–48),

> Violence provokes things. You start to expect it. You learn to inhabit your body in a different way with that expectation. When you perceive the outer world as a danger, what changes is your relationship to your body: you become more cautious, timid (…). They tell you to be careful: that you should take all possible precautions and you anxiously await the possibility of their destroying you. You start to learn that being careful, trying to avoid these things happening to you, is a way of avoiding their hurting you (…). Becoming a young woman is to learn to expect these insinuations, to alter your behaviour as a result; to be cautious because you are in a public space; to be cautious simply because of being a young woman. In fact, if you don't alter your behaviour correspondingly, if you aren't careful and cautious, they can make you responsible for the violence that they have inflicted on you (look what you were drinking, look how you were dressed, look where you were, look, look).

Violence is turned into a lesson when it is accompanied by a narrative, an explanation (Ahmed 2018). One paradigmatic example from the early 1990s is the narrative on sexual danger of the so-called *Alcàsser case*. According to the official version, three teenage women from this town where hitchhiking and got into a car occupied by at least two men to go to a discotheque in a nearby locality. Three months later, their corpses were found with marks of sexual torture (Barjola 2018, 17). For Barjola (2018, 29), the narrative created around this case was essentially "an instructive warning" for a whole generation of women who were starting to take over the public space, incorporating feminist principles into everyday life. In Spain, the 1980s was a period of considerable feminist activity that crystallised in the partial depenalisation of abortion, the divorce law and the reform of the penal code on sexual matters, which eroded the axes of the patriarchy in relation to the family institution and the control of women's bodies and sexuality (Barjola, 2018, 64). Barjola (2018, 64–67) considers the narrative on the forced disappearance, subjection to sexual violence and murder of those three teenagers to be a technology of control aimed at alleviating the advances made by the feminist movement "in redefining women's bodies and men's rights over them". To that end, the narrative uses the enormous impact of the crime to focus attention on its horrifying aspect, shifting attention away from its causes and placing the teenagers' behaviour at the centre of the explanation – going out at night and hitchhiking as transgressive of social norms – thus making them responsible for what happened[1] and absolving the aggressors (Barjola 2018, 272–273).

Decades later, sexual violence is produced and normalised in a scenario that largely responds to recent social transformations that give rise to new patterns of carrying out aggressions and include the public agora that is now formed by social media (on which contacts are made and aggressions against women are visualised)[2] and the massified leisure spaces. These include popular festivals in neighbourhoods, towns and cities, as well as discotheques or places with a large number of bars where people socialise, in which there are high levels of alcohol consumption starting from a very young age (Rubio & Sanz-Díez de Ulzurrun 2018).

The festivals of each locality, along with forming massive leisure areas, occupy a central place in the community's life and relational framework. These festive spaces and their cultural codes are not alien to the gender systems and powers that underlie the constant threat that women must live with and the fear memorised in their bodies, which restricts their possibilities of enjoyment (Guilló 2016). However, the body is not only a site of discrimination but also of resistance and opposition to social structures (Esteban 2004). Facing sexual and sexist aggressions and the reproduction of gender, feminist actions have acquired centrality and shaped the festivals as a political space of assertion, which demands collective pleasure linked to the well-being of women (Guilló 2016).[3]

Legislative framework, institutional and social feminism

In the Autonomous Community of the Basque Country (ACBC) there is an institutional framework focussed on constructing real and effective equality between women and men. Its origin lies in the 1980s, in what is known as institutional feminism that in Spain was related to the Spanish Socialist Workers' Party taking office in 1982. The Spanish Women's Institute was created in 1984, while in the ACBC, Emakunde-Basque Women's Institute was formed in 1988 as an autonomous public body directly dependent on the presidency of the Basque government. Emakunde was commissioned to develop, write and bring to the Basque Parliament Act 4/2005, dated February 18, for the Equality of

Women and Men, which addressed questions related to violence against women. Concretely, Act 4/2005 understood violence against women to be

> any violent act based on sex that results, or might result, in physical, sexual or psychological harm or in the suffering of women, including threats to realize such acts, coercion or the arbitrary privation of liberty that might occur in public or private life.
>
> (Article 50)

This definition includes aggressions suffered by women in festival settings; however, the act (prior to its modification, currently underway) does not specify questions relating to festivals beyond what is stipulated in Article 25 on parity in women's participation in festivals (the article was drawn up ad hoc to cover the *Alardes* (parades) in Irún and Hondarribia), forbidding Basque public administrations from granting funds and participating in cultural activities and festivals when equality is not guaranteed.

At the state level, we can underscore two organic acts promoted by the socialist government of José Luis Rodríguez Zapatero: Organic Act 3/2007, dated March 22, for the effective equality of women and men (henceforth the Equality Act) and Organic Act 1/2004, dated December 28, on Measures for Integral Protection against Gender Violence (henceforth the Gender Violence Act), which had pioneering character in Spain and Europe.

The Equality Act does not make any express mention of gender violence occurring in festival settings, but Article 7 of the act defines sexual harassment and harassment based on sex as "any behaviour, verbal or physical, of a sexual nature that has the aim or effect of attacking a person's dignity, in particular when an intimidating, degrading or offensive environment is created" (Article 7.2) and harassment based on sex as "any behaviour based on a person's sex with the aim or effect of attacking their dignity and of creating an intimidating, degrading or offensive environment". Further on, Article 14, on establishing the criteria for general action by the public powers, makes them responsible for adopting "the necessary measures for the eradication of gender violence, family violence and all forms of sexual harassment and harassment based on sex" (Article 14.5). However, the measures to prevent sexual harassment and harassment based on sex are circumscribed to settings of labour (Article 48) and the public administrations (Article 62).

For its part, the Gender Violence Act of 2004 states in the exposition of its motives that "a technical definition of the syndrome of the maltreated women already exists", which consists in

> the aggressions suffered by a woman as a result of sociocultural conditioning factors that act on the male and female genders, placing her in a position of subordination to the man and manifested in the three basic fields of personal relations: maltreatment in the relations of the couple; sexual aggression in social life; and harassment in the labour setting.

The first article states, "The gender violence to which this present Act refers encompasses every act of physical and psychological violence, including aggressions against sexual freedom, threats, coercions or the arbitrary privation of liberty" (Article 1.3). Nonetheless, despite its defining gender violence as that suffered by women for the fact of being women, and in spite of including sexual aggression in social life as an expression of this, the act does not specify these questions given that Article 44 restricts those who are to be considered victims of gender violence: only those women who suffer violence from men with whom they have, or have had, a relation of 'affectivity'. This specification can be considered a limitation, above all if we refer to the violence women suffer in festival settings, because, in the majority of cases, that sexual violence, those sexual abuses and aggressions, proceed from men who are strangers. In fact, during the San Fermín Festival of 2008 in Pamplona, Nagore Laffage was murdered by Diego Yllanes, and, during the trial, the Gender Violence Act was not applied because it was considered that, in spite of the murderer and the victim knowing each other (they worked in the same hospital), there was no romantic relationship between them. Sexual abuses and aggressions are included in Act 35/1995, dated December 11, of Aid and Assistance to the Victims of Violent Crimes and Against Sexual Freedom.

Returning to the ACBC, and prior to the Equality Act, we must underscore the Interinstitutional Agreement to improve the care provided to women who are victims of domestic maltreatment and sexual aggressions. This act was promoted and coordinated by Emakunde in 2001 and aimed to

> guarantee to women suffering this type of violence the most integral and coordinated care possible in the health, police, judicial and social aspects. To that end the patterns and criteria of action to be followed by professionals in such cases are established, and stable channels for cooperation amongst the institutions concerned are assembled
>
> (p. 9)

It was signed by the president of the ACBC and the representatives of the General Council of the Judiciary, the Public Prosecutor's Office of the High Court of the Basque Country, the three Provincial Councils (Alava, Biscay and Gipuzkoa), the Association of Basque Municipalities (EUDEL), the Council of Lawyers of the Basque Country and the Basque Medical Council. The re-edition of the agreement in 2011 was also signed by the security forces of the state and the autonomous community. The agreement establishes a protocol for action in the health, judicial and police fields, as well as the social services and law societies. The protocols focus on care for the victims of maltreatment or sexual aggressions, favouring care that is personalised and informed and, above all, coordinated interinstitutionally. This first protocol did not yet contain actions with a preventive character.

Primary and secondary preventive measures were to be promoted above all from the municipal field. Actions to be taken to prevent gender violence in festival settings were not to be isolated and unconnected. The prior institutional framework was important for generating awareness that gender violence is a public and social problem. From an institutional point of view, the creation of Berdinsarea (Network of Basque Municipalities for Equality and against Violence to Women) signed by EUDEL and Emakunde – derived from the Interinstitutional Agreement – was very important in promoting and encouraging the design and establishment of measures of preventive action for cases of gender violence in festivals in towns and cities. The basic document of Berdinsarea, signed in February 2005, defines its mission as,

> encouraging, strengthening, coordinating and evaluating programs and services managed by the local administrations in favour of equality and against violence to women, by means of defining joint criteria for intervention and evaluation, framing this work in an integral intervention that permeates all levels of local public action and utilizes the Network as a nexus of union between this and other supra-municipal bodies.
>
> (p. 2)

Its principles contain an integral view of gender violence and prioritise its prevention. Amongst its aims, it is worth underscoring one that has had a de facto impact on the promotion of concrete interventions:

> To promote the adoption of agreements of collaboration and protocols for intervention that develop, concretize and adapt to the municipal reality the content of agreements adopted at the supra-municipal level relating to equality and the fight against violence to women.
>
> (p. 3)

The effectiveness of the actions promoted by the Berdinsarea network are also related to the fact that one condition for joining this network is the fulfilment of certain prior requirements that are what, a priori, will make it possible to design, implement and evaluate measures of equality and prevention against violence to women. These prior requirements are (1) having an equality plan or program for women and men (as demanded by Act 4/2005); (2) availability of technical personnel with experience and/or specific training in questions of equality between men and women to promote that equality plan or program; (3) guaranteeing sufficient resources in the council budget; (4) relations and channels for participation and collaboration with women's associations and other organisations that contribute to attaining equality of women and men in the local fields; (5) complying with the requirements of current legislation on questions of prevention, care and protection of victims of domestic maltreatment and sexual aggressions; and (6) the commitment to take an active part in the network.

The re-edition of the agreement in 2004 concretised a series of interventions and commitments that include creating a structure (technical secretariat), providing economic resources (a budget allocation of over 50,000 euros by Emakunde) and prioritising training activities aimed at the political and/or technical personnel of the local administration in the field of equality and violence to women, together with the creation of inter-municipal working groups to address issues related to both questions.

Berdinsarea has published a series of documents that not only aim to encourage but also orientate local government activities. In 2014, a series of recommendations were published on public interventions facing cases of violence against women, two of which stand out: (1) recommendations for relating to the victim and her environment in cases of sexual aggressions or other manifestations of violence against women and (2) recommendations for convening demonstrations in cases of sexual aggression and other manifestations of violence against women. These recommendations completed the existing protocol on how to act in cases of gender violence resulting in death. In this case, local political powers were recommended

> to advance in the agreement with the associative fabric to convene joint actions rejecting violence against women, understanding that this joint intervention is the better course, but that both institutions and associations have the right to convene the population when they consider this necessary.
>
> (p. 21)

Such rallies not only aim to break the silence surrounding such events but also to raise awareness and understanding. The majority of the municipalities integrated in Berdinsarea, and those that do not form part of it, have adopted this recommendation and the need to hold rallies has spread, not only in cases of the murders of women but also in cases of sexual aggressions. In the majority of cases, these mobilisations are jointly convened with the feminist movement. A clear example of the normalisation of this social and political response is provided by the case of the Wolf Pack: the day after the gang rape, there was enormous social participation in the rally convened by Pamplona City Council on July 8, 2016.

Besides rallies against sexual aggressions, several Basque town councils have implemented a series of preventive actions during the celebration of their festivals (information stands, pamphlets, campaigns, symbols). Many of these measures favour awareness raising since they stress the consideration that violence against women during festivals is a social problem in which alcohol, drugs and the festive ambience cease to be considered justificatory elements and become mere disinhibitors of behaviour arising from patriarchal and sexist structures. Nonetheless, other measures, such as services to accompany women home during the festivals in some municipalities, serve to strengthen gender stereotypes, women's fears and their lack of freedom to enjoy the public space.

We cannot end this section without mentioning that all the legislative advances and the local measures of Basque institutional feminism are not alien to the role played by the Basque feminist movement. According to Eva Martínez, "the appearance of more transformative groups, the emergence of doubts and the questioning of the feminist subject must be situated, with some exceptions, at the start of the new century" (Martínez 2019, 61). She identifies the IV Feminist

Conference of 2008 and the State Conference of 2009 as turning points. Nonetheless, in the 1970s and 1980s, violence against women was one of the issues that brought together the different assembly-based movements of that time: "Denunciation was mainly focused on sexual aggressions occurring in the public sphere" (Martínez 2019, 112). At the start of the twenty-first century, the World Women's March and the publication of the World Charter of Women, which had emerged in Canada to demand economic justice (Díaz 2017), also spearheaded the fight against violence to women in the Basque Country. Returning to Eva Martínez (2019, 162), we can state that one of the main contributions of the World Women's March was the protocol in relation to women murdered due to sexist violence, by which feminist groups committed themselves to holding rallies when a death occurred in such circumstances. The Basque feminist movement has led these rallies, just as it has enabled rallies to be organised on November 25 or March 8, or one-off events when legislative proposals arise on questions of equality or gender violence.

Methodology

A qualitative methodology was used to analyse the experiences of young women in festival settings, as it enables us to enquire more deeply into subjective reality, meaning, lived experiences, perceptions, attitudes and cultural codes (Ruiz Olabuénaga 2012). The following are the specific objectives guiding the applied part of this work:

1 Analyse political and social interventions by the feminist movement and institutional feminism.
2 Collect lived experiences of young women in the festival setting of the Basque Country, with special mention of sexual and sexist aggressions, whether experienced directly or heard about.
3 Identify the incorporation of mechanisms of feminism and empowerment by young women.
4 Explore through the discourses of young women the scope and perception of the feminist interventions that have been realised.

We held two focus groups with 12 young women (aged between 20 and 22 years). The technique of the focus groups and the analysis of the results were based on content analysis. This enabled us to enter into the lived experiences and collective discourses of these young women and contrast their perceptions and lived experiences with our prior theoretical framework, the legislative measures and, above all, the policies implemented by the Berdinsarea network in the ACBC.

Analysis of the lived experiences of the young women

The festivals in towns and cities occupy an important and central place in the lives of the young women who participated in the focus groups. In fact, as they state, their significance is more transcendental than material and, according to the young women, they are spaces that enable one "to feel part of" one's province, town or city. Popular festivals have significant particularities in the cultural imaginary of the Basque Country. Amongst other questions, they occupy the public space in a highly significant way, and their organisation to a large extent involves citizen collaboration, neighbourhood and hostelry associations, collectives, etc. Without any question, participation in these festivals involves a feeling of belonging.

Although such festivals and events are of crucial importance to the young women, the scenario they describe reproduces gender systems and power schemes. Furthermore, all of them clearly understand that in such contexts, it is not the same thing to be a young woman or a young man – with the latter playing a central role in the festivity. In their own words:

(NEREA, 21 years old): When something has to be announced, or someone has to get up on a stage, it's always guys who do it. They are usually the protagonists.
(LAURA, 22 years old): For example, the bands that come to play, or the DJs. They're all guys, six DJs come, and they're all guys or the main bands.
(LORENA, 22 years old): They arrive at a time when there are only guys. They invade absolutely everything, and, what's worse, this happens when there are the most activities.
(IRATI, 21 years): In the end, it's the guys who make the most advances; they continue to have the initiative. For example, it's absolutely normal for a guy to invite a girl for a drink, but if a girl does that to a guy, it's badly seen. For my part, I've never seen it happen.

Nonetheless, one of the most oppressive mechanisms detected in the discourses of all the young women is the culture of fear that impregnates their bodies and their discourses. Very diverse socialisation agents, the mass media, their families and so forth have taught them to live with fear in the streets and in the festivals. As they themselves tell it:

(NAIA, 21 years): When you start to hear about cases happening close to you, then you start to doubt everything because something that seemed safe to you is no longer so safe.
(USUE, 22 years): We have to face constant questioning at home about who we came back with, what time we got back.

These fears and anxieties, impregnated in their memories and bodies, mean that they experience the festivals "on the alert". They do not feel free to enjoy them; they feel inhibited and have to hold themselves back, especially when it is time to go home, which is when they feel most vulnerable. The following statements refer to this:

(LORENA, 22 years): At the main entrance to where I live, when I get there, I always try to hurry in and close both doors because I think there might be a guy between the two doors; it's a worry that's always on my mind.

(ITZIAR, 21 years): Getting robbed is the worst that can happen to a guy, but for us, that's the least of our worries. You have to be constantly on the alert.

(USUE, 22 years): You always live with fear, even though you feel you're being foolish. But it seems that you aren't able to defend yourself on your own; it's like you're oppressed.

In addition to the culture of fear, the reality is that all the young women who took part have had to face some kind of comment or (hetero)sexist attitude and some type of sexual aggression. The comments that follow are examples:

(LAURA 22): When you're in a bar they treat you like dirt; you always have to be on the alert. You're alert because the first thing you think is that they're going to harass you, that they'll invite you to have a drink, that they'll call you "brunette", "blonde". Fuck, I'm a person.

(MARTA, 22 years): Once, on the way home with a friend, a man started to follow us and shout things like "you're hot", "I'd like to screw you", "what an ass!" We went inside the main entrance, and the guy waited outside for my friend to come out, and her father had to come and fetch her.

(NEREA, 21 years): Once I was wearing a skirt, and a guy took out his penis and started rubbing it against me; it was really disgusting!

Similarly, many young women relate that if they want to enjoy their sexuality in festival settings, they have to face the stigma of being called a "whore". That means that besides being on the alert, they feel unsafe, fearful and coerced. All of this creates a feeling of impotence but also guilt at not knowing how to respond and act in the face of aggressions. As can be seen in the following extracts, the culture of fear and systemic oppression, exercised in their memories and bodies, has taught them to feel defenceless. They have learned not to respond to these aggressions:

(LORENA, 22 years): You don't react, and what's worse, you blame yourself for not responding, for not shouting, for not reacting.

(MARTA, 22 years): You feel blocked; you don't know how to answer back, and even in such cases, you feel afraid to respond. It's like you can't defend yourself on your own.

The young women's perceptions of the involvement of young men in this type of aggression are highly significant. They mention how fed up they are with the indifference of young men in the face of such conduct, which they regard as normal. They talk of a lack of understanding and empathy, and they even relate how these aggressions at times become a mechanism for validating masculinity. As recounted by Laura, "There are many guys who encourage each other to carry out these aggressions. I swear they behave like monkeys.... Like: 'look, look, she's ready for it!'"

For all these reasons, the young women feel that their female friends are their only source of support, as well as their families. Amongst themselves, they have generated strategies and codes of "salvation" facing sexist and sexual aggressions. The following are a few examples:

(LAURA, 22 years): Guys come up making a nuisance of themselves, and you have to make signals to your friends to come and rescue you.
(IZASKUN, 21 years): In our group of friends, we have a secret sign, which is touching one's ear, and if we see that one of our group is with a guy and making the sign, we go and rescue her.
(IRENE, 22 years): A guy who was much taller than me started to back me up against the wall; he had me hemmed in, and a [female] friend appeared and gave me a kiss, so the guy left.

The response to these types of abuses, aggressions and male chauvinist and sexist situations has drawn sustenance from feminist content and arguments. In the discourses of our informants, we find a positive evaluation of the actions that the feminist movement is carrying out. The actions they mention include banging pots and pans or stopping the music in response to possible sexual and sexist aggressions, celebrating 'feminist days', stopping male chauvinist music, providing more portable toilets, posters and feminist 'merchandising'. Nonetheless, although they evaluate these interventions positively, all the young women who participated in the focus groups say that they do not identify with, belong to or feel linked to these spaces.

In general terms, they believe structural changes are still needed. Very few of the informants mentioned actions carried out by the institutions; they mention some posters and the operations to escort women home organised by some municipalities, like in Santurce. However, they do not feel that the actions respond to a real social commitment, and they state that important challenges remain that require more work.

Conclusions

The opinions we obtained question the power structure and denounce the inequality between women and men in the occupation and use of the public space during the popular festivals, a space where they have to face attacks on their bodies, on their sexualities and on themselves. Bodies become territories of memory, and the women are afraid. It is a fear that paralyses them, blames them and makes them feel insecure, which is caused by their socialisation in a culture of fear that diminishes and subordinates them.

The role played by men in festival settings proves to be an important issue that must be addressed, given that, according to our testimonies, an absent role is perceived, which positively encourages sexist and male chauvinist behaviours. It seems that young men have normalised and neutralised sexual aggressions and abuses, while in spite of this, preventive campaigns continue to focus on women as possible victims. It is important to socialise the questioning of male privileges.

The fight of the feminist movement is undisputed and recognised. Multiple actions have been generated that are producing a tacit, symbolic and practical transformation. Nonetheless, despite the mainstream character of feminist affirmations and the concretisation and closeness of the political subject, it is still something alien to, or remote from, many young women.

The measures promoted by institutional feminism, both juridical and political, have generated fertile ground that enables advances to be made in socialising the feminist discourse in festival settings. However, our young informants continue to point to challenges that are pending, such as giving greater visibility to the actions undertaken, more areas of interlocution and greater scope of feminism in politics to be able to tackle the problems at their roots.

Violence against women is a socialised conduct, a social construction in a broad sense; that is, it forms part of the process of socialisation and is transmitted by socialising agents – family, school, peer groups and mass media (Rubio & Sanz-Díez de Ulzurrun 2018 52). Creating the conditions for its elimination involves transforming and resignifying bodies, spaces and individual and collective practices. Central to all of this is the agency and re-appropriation of women's bodies (Lagarde 2012, 244), their shift from objects to be possessed to subjects with the possibility of resisting and providing a different – non-patriarchal – interpretation of reality (Esteban 2004, 63).

In recent years, the alliance of women, feminist rallies and mobilisations – closely related to the Wolf Pack case in 2018 and 2019, #MeToo, and the use of social media – transcended the women's alliances concerned with violence against women that were analysed by Arnold (1995). We are perhaps witnessing a new wave of feminism that is producing a serious reaction, just as de Beauvoir (2000) and Susan Faludi (1991) predicted. Furthermore, it is also having a clear effect on politics and the institutions, as shown by the ruling of the Spanish Supreme Court in the case of the Wolf Pack, which contains assertions that any feminist would support.

Notes

1 Gil (2008) and Vigarello (1998) provide documentary evidence that questioning the behaviour of victims of sexual aggression in Spain and France, respectively, dates back centuries.
2 The technological, social and cultural advances brought by the Internet run parallel to the generation of online risks marked by the traditional conditioning that proceeds from the offline world, such as the patriarchal system and sexist culture (Zafra, 2010). In this way, different forms of online abuse are generated, which produce spaces of intimidation against women's bodies and sexuality (Powell & Henry 2014; Navarro, 2016; EIGE, 2018).

3 From a different perspective, Esteban (2016, 99) refers to the shared use made by the members of a feminist group of the time of preparing to go out (at the weekend to festivals) as a space/time for entertainment and communication, for affirming their diverse and alternative identity and for criticism and self-criticism: "a ritual of sociability and preparation for their integration as equals".

References

Ahmed, S. (2018). *Vivir una vida feminista (Living a feminist life)*. Barcelona: Edicions Bellaterra.

Arnold, G. (1995). Dilemmas of feminist coalitions: Collective identity and strategic effectiveness in the battered women's movement. In M. M. Ferree (Ed.), *Feminist Organizations: Harvest of the New Women's Movement* (pp. 276–290). Philadelphia: Temple University Press.

Barjola, N. (2018). *Microfísica sexista del poder. El caso Alcàsser y la construcción del terror sexual (Sexist microphysics of power: The Alcàsser case and the construction of sexual terror)*. Barcelona: Virus.

Basaglia, F. (1983). *Mujer, locura y sociedad (Women, madness and society)*. Puebla: Universidad Autónoma de Puebla.

Brownmiller, S. (1975). *Against our will: Men, women and rape*. New York: Bantam.

Connell, R. W. (1995). *Masculinities*. Berkeley: University of California Press.

De Beauvoir, S. (2000). *Le deuxième sexe (The second sex)*. Paris: Gallimard.

Díaz, C. L. (2017). La Marcha Mundial de Mujeres. Feminismos Transnacionales en movimiento (*Global Woman protest. Transnational Feminisms in movement*) (Thesis Dissertation). https://ciesas.repositorioinstitucional.mx/jspui/bitstream/1015/469/1/TE%20 D.A.%202017% 20Carmen%20Leticia%20Diaz%20Alba.pdf. Accessed 14 Nov 2019.

Esteban, M. L. (2004). *Antropología del cuerpo. Género, itinerarios corporales, identidad y cambio (Anthropology of the body: Gender, bodily itineraries, identity and change)*. Barcelona: Bellaterra.

Esteban, M. L. (2016). Cuerpo e imagen corporal: cambios, rupturas e hipersexualización femenina (Body and bodily image: Changes, ruptures and female hyper-sexualization). In M. L. Esteban, M. Bullen, C. Díez, J. M. Hernández, & E. Imaz (Eds.), *Continuidades, conflictos y rupturas frente a la desigualdad: Jóvenes y relaciones de género en el País Vasco (Continuities, conflicts and ruptures in face of inequality: Young people and gender relations in the Basque Country)* (pp. 89–99). Vitoria-Gasteiz: Emakunde-Instituto Vasco de la Mujer.

European Institute for Gender Equality (EIGE). (2018). *Gender equality and youth: The opportunities and risks of digitalisation*. Lithuania: EIGE.

Faludi, S. (1991). *Reacción. La guerra no declarada contra la mujer moderna (Backlash: The Undeclared War against American Women)*. Barcelona: Anagrama.

Giddens, A. (2000). *La transformación de la intimidad en las sociedades modernas (The transformation of intimacy in modern societies)*. Madrid: Cátedra.

Gil, A. (2008). *Historia de la violencia contra las mujeres. Misoginia y conflicto matrimonial en España (History of sexual violence against women: Misogyny and matrimonial conflict in Spain)*. Madrid: Cátedra.

Guilló, M. (2016). *Festak, genero-harremanak eta feminismoa (Festivals, gender relations and feminism)*. Bilbao: Udako Euskal Unibertsitatea.

Lagarde, M. (2011). *Los cautiverios de las mujeres (Women's experiences of captivity)*. Madrid: Horas y HORAS.

Lagarde, M. (2012). *El feminismo en mi vida. Hitos, claves y topías (Feminism in my life: Landmarks, keys and utopias)*. México: Inmujeres DF.

Martínez, E. (2019). *Redes de alianza y coordinación en el movimiento feminista vasco (1975–2014) (Bond and coordination networks in the Basque Feminism movement. (1975–2014))* (Thesis Dissertation). Universidad del País Vasco.

Navarro, R. (2016). Gender issues and cyberbullying in children and adolescents: From gender differences to gender identity measures. In R. Navarrro, S. Yubero & E. Larrañaga (Eds.), *Cyberbullying Across the Globe. Gender, Family, and Mental Health* (pp. 35–44). Cuenca, España: UCLM.

Observatorio Noctámbul@s. (2018). *Observatorio sobre la relación entre el consumo de drogas y las violencias sexuales en contextos de ocio nocturno (Observatory on the relationship between drug consumption and sexual violence in contexts of nocturnal leisure)*, 5º Informe anual 2017–2018 (5th Annual Report 2017–2018). Barcelona: Fundación Salud y Comunidad.

Osborne, R. (2009). *Apuntes sobre violencia de género (Notes on gender violence)*. Barcelona: Edicions Bellaterra.

Powell, A., & Henry, N. (2014). Blurred lines? Responding to "sexting" and gender-based violence among young people. *Children Australia, 39*, 119–124. doi:10.1017/cha.2014.9

Puleo, A. H. (2005). Lo personal es político: El surgimiento del feminismo radical (The personal is political: The emergence of radical feminism). In C. Amorós & A. De Miguel (Eds.), *Teoría feminista: de la ilustración a la globalización. Del feminismo liberal a la posmodernidad (Feminist theory: From the Enlightenment to globalisation: From liberal feminism to postmodernity)* (pp. 35–67). Madrid: Minerva Ediciones.

Rubio, A., & Sanz-Díez de Ulzurrun, M. C. (2018). Violencia sexual contra las mujeres jóvenes: Construcción social y autoprotección (Sexual violence against young women: Social construction and self-protection). *Revista de estudios de juventud, 120*, 47–65.

Ruiz Olabuénaga, J. I. (2012). *Metodología de la investigación cualitativa (Methodology of qualitative research)*. Bilbao: Universidad de Deusto.

Segato, R. L. (2003). *Las estructuras elementales de la violencia. Ensayos sobre género entre la antropología, el psicoanálisis y los derechos humanos (The basic structures of violence: Essays on gender between anthropology, psychoanalysis and human rights)*. Buenos Aires: Universidad Nacional de Quilmes.

Sortzen. (2011). *Agresiones sexuales. Cómo se viven, cómo se entienden y cómo se atienden (Sexual aggressions: How they are experienced, how they are understood and how they are treated)*. Vitoria-Gasteiz: Gobierno Vasco.

Vigarello, G. (1998). *Historia de la Violación. Siglos XVI-XX (History of rape: XVI–XX centuries)*. Madrid: Cátedra.

Zafra, R. (2010). *Un cuarto propio conectado. (Ciber)espacio (auto)gestión del yo (An online room of one's own: (Cyber)space and (self)management of the I)*. Madrid: Fórcola.

3 *Fiestas*, public space and rape culture. A study of the Wolf Pack[1] case

*Anna Morero Beltrán and
Clara Camps Calvet*

Introduction

On July 7, 2016, during the festival of San Fermín in the city of Iruña,[2] a group of men committed an act of sexual aggression. Because of the characteristics of the attack and the social and political climate with regard to sexist violence in Spain, the event came to mark a turning point in the country's responses to sexual violence in general and, more specifically, to attacks perpetrated in public space during leisure activities and celebrations. A young woman, 18 years of age, was sexually assaulted by a group of five men who referred to themselves as 'the Wolf Pack'. The woman was on her way to the car from the celebration to rest when she came across this group on the street, and they offered to accompany her. Later, they attacked her in the entrance hall of an apartment building. The men were arrested just a few hours after the attack and a few days later were ordered to report to jail. The courts found that the crime they had committed could not be labelled a sexual assault, saying that the attack had not involved violence or intimidation. Instead, they categorised the crime as an act of sexual abuse. The verdict was met with strong opposition from the feminist movement in Spain, with street demonstrations throughout the country. The protesters maintained that the decision was evidence that the justice system was founded upon the logic of patriarchy: a framework that reproduces the myths and stereotypes that reinforce rape culture (Brownmiller, 1976).

For various reasons, the case came to be seen as paradigmatic. Firstly, San Fermín is a very popular traditional outdoor festival in Spain, a celebration that has garnered growing international renown. However, the festival has also been the location of a number of sexual attacks and of a femicide. In 2008, Nagore Lafagge was murdered as she tried to escape her rapist. Like in the case that is the focus of this chapter, the episode began in a public space, the first contact with the aggressor apparently occurring with the victim's consent. When the two were in the aggressor's house, however, Nagore was raped and murdered after refusing to have sexual relations. Another form of sexual violence occurs during the Txupinazo,[3] an event that marks the beginning of the San Fermín celebrations and attracts thousands of people to the square outside Pamplona's city hall. The presence of this crowd tends to facilitate sexual violence, as women are often publicly harassed by men with the approval of onlookers. These aggressions are only a few of the most visible examples, prominent enough to attract the attention of the media. Women who attend San Fermín often view sexual aggressions as an implicit part of the event, something they know they will have to face when they are present in leisure spaces.

Secondly, the case has helped to place sexual violence at the heart of the public debate and to shed light on the links between this violence and the broader patriarchal society. The fact that the aggression was planned in advance and perpetrated by a group of five men (four of whom are also pending trial for another group sexual assault) allowed the feminist movement to show that rape has nothing to do with intimacy or desire. On the contrary, the aim of sexual violence is power and control, reinforcing a patriarchal conception that elevates men's understanding of sexuality and totally discounts women's experiences (MacKinnon, 2018). In fact, this case is evidence that sexual violence against women is connected to what Rita Segato (2003, 2016) has called the "masculine mandate": an obligation placed on men to engage in constant displays of strength and virility before their community of peers. This display of masculinity often takes the form of the subjugation of women's bodies, an integral and necessary element of the set of practices that reproduce hierarchical power structures and reinforce gender status. It was very clear that the sexual aggression in this case was a demonstration of masculinity. The aggressors shared a number of photos and videos of the attack with other men via two WhatsApp groups, accompanied by messages such as, "The five of us are all fucking this one girl", and "What a fucking amazing trip!" Even more revealing was a WhatsApp exchange that occurred a few days before the group's trip to Iruñea, when a member of the group wrote, "This vacation is the true test of whether you are a wolf". Meanwhile, one of the aggressors had a tattoo reading, "The wolf's power comes from the pack". The power of the wolf, the man suggested, lay in his ability to demonstrate his status as a "wolf" to other men by subjugating a woman's body in an act of sexual violence.

Thirdly, the feminist movement responded very forcefully in the streets to the sexual aggression. Although feminists have long been striving to eradicate sexist violence, the reaction to this case represented a turning point in this struggle in Spain, helping to shape a new framework of interpretation of sexist violence in general and sexual violence in particular. This new context forced the public administration to respond (albeit insufficiently) to the violence perpetrated in this sort of setting – a problem that governments had taken no specific steps to address prior to these events.

The chapter will show that the events that occur at these *fiestas* are not the product of isolated situations far removed from the logic of broader society. Instead, they are representations and reproductions of the dominant social order. In other words, *fiestas* function not as mere leisure activities but as practices that both represent and legitimise fundamental aspects of our culture and society, serving especially to reinforce dominant values, models and hierarchies (Farapi, 2009, 207). Our analysis of the 'Wolf Pack' case will examine men and women's behaviour and the codes that define the relations between them in public spaces during festive events (and in liminal spaces between the public and private realms). The chapter will show how these behaviours and codes are exploited by those who would argue that sexual relations occurring in private spaces are consensual. This chapter will apply its theoretical framework to the judicial record of the 'Wolf Pack' case to show how the behaviour of women in public spaces is often placed at the centre of legal analyses, with courts taking these behaviours as supposed proof of victims' consent to sexual relations and as evidence to exonerate perpetrators of sexual aggressions. We argue that a patriarchal logic tends to prevail in the uses of public space, especially when it comes to outdoor celebrations. This mentality also permeates the judicial system, which legitimises an idea of consent that does not require renewed consent for sexual activity. *Fiestas* tend to be depicted as neutral spaces for enjoyment, events at which the normal rules that govern our everyday lives and behaviours are temporarily suspended, and transgressive behaviour is allowed. However, we conclude that these spaces remain social mechanisms that promote a vision of consent that is based on the logic of patriarchy, one that contributes to reproducing rape culture. Finally, the chapter discusses how sexual violence functions to dispossess women of their claim to the right to public space and how the feminist movement has fought to redefine concepts in such a way as to reclaim this space.

This chapter will contribute to the building of theoretical bridges between the analysis of space from a feminist perspective and other theoretical work on sexual violence. In this analysis, the concept of consent is the key to understanding how sexual violence in public and private space is structured, configured and expressed.

This chapter consists of five sections. The first section highlights some of the most important theoretical contributions dealing with the configuration of space and of the sites of public *fiestas* in particular. The second section explains the methodology used to carry out this study. The third section places special emphasis on the phenomenon of sexual violence in Spain and the festival of San Fermín. There is also a discussion of the feminist movement's response to the case. The fourth section is based on the contributions of interviewees and a content analysis of the court's decisions in the case. The section features an in-depth analysis of how the patriarchal logic that dominates the uses of public space for festivities ultimately becomes embedded in judicial decisions on consent in cases of sexual assault. These results make it possible to sketch the connections between the case and the theoretical contributions discussed in the first section. Finally, the fifth section presents the conclusions of this study.

Theoretical framework: public space, *fiestas* and sexual violence[4]

Scholars in the field of feminist geography have shown how cities and the spaces within them are gendered. Consequently, women are often excluded from public space, or their presence is allowed only as long as they occupy certain predetermined, limited roles (Ruddick, 1996 quoted in Rodó de Zárate & Estivill i Castany, 2016). When we apply this notion to the context of public *fiestas*, it is clear that these events are manifestations of the social order, spaces dominated by certain generally accepted values, models and power structures. In addition to representing the social order, these celebrations also function to legitimise these structures via ritualised activities featuring the participation of the community and of a range of representatives of power (Farapi, 2009).

Heteropatriarchy establishes a set of roles for each gender and exerts discipline over bodies to enforce these rules and to ensure that people behave in certain ways in public space. For example, the very fact that women are visible in the street has implications for how they experience public space, as their bodies are sexualised by the male gaze (Hyam, 2003). This inevitably effects how women experience the city and their leisure activities. Far from serving as a break with the structures of everyday life, *fiestas* reproduce the dominant sociocultural representations and practices in concentrated form, accentuating the conflicts and the discrimination that exist in everyday society. For example, the atmosphere of these celebrations reinforces gender roles, power relations, exclusion of certain groups, masculine frame of reference and reproduction of violence (Zuloaga et al., 2018).

It is important to emphasise how space can function to cement social differences and hierarchies. This spatial segregation is linked to social segregation along gender lines, as "the spaces that women occupy, and those that are off limits to them, are closely connected to the social order of a given place and time" (Valle, 1997, 35). The sexual differences that have shaped our societies have also structured our cities as places where the two sexes do not have equal rights to the use of space, with supposedly neutral space in fact marked by this sexual inequality (Patiño-Díe, 2016).

It is patriarchy that has made the city into a place by and for men, where women's presence is an 'intrusion'. This is palpable in the insecurity of certain spaces or in the very way people relate to one another, with women on the street alone interpreted as being sexually available and as lacking any right to privacy, when a woman walking down the street is a walking display that invites catcalls, whistles, stares (Rodó de Zárate & Estivill i Castany, 2016) and even attacks.

Further, stereotypes, the assigning of gender roles, the conception of bodies and the perception of danger are all conditioned by the structures of heteropatriarchy, and this structure shapes gendered relations in space (Rodó de Zárate & Estivill i Castany, 2016). The perception of fear and the resulting spatial behaviours depend on age, abilities and place of origin. Regardless of the differences in their social identities, however, women feel this fear, as fear precedes or goes hand-in-hand with violence (Falú, 2009).

Thus sexual violence is best understood as a foundational building block of patriarchy, given that any assault is not just an attack on one individual woman. On the contrary, all women feel the impact of these attacks, as this sexual terror deprives women of their ability to use and enjoy public space (Barjola, 2018). This terror produces a kind of bodily knowledge, one that begins from a very early age. Thus sexual violence is incarnated gradually yet constantly (Barjola, 2018). According to Velázquez (2003), the geography of fear functions through physical and psychological self-censorship, and it acts as a tool to force women to internalise the mandates of gender and to limit how they exercise their freedom.

That is why, as Barjola (2018) pointed out, it is critical to immerse ourselves in counter-representations in order to undermine this message of sexual danger. This production of 'counter-representations' provides us with a new framework, a new reading that leads directly to the feminist empowerment of women as they adopt a strategy of direct confrontation of terror and sexual violence.

Methodology

A range of methods was used in this research. In-depth, semi-structured interviews were conducted with a number of people involved in the festival of San Fermín: members of self-organised feminist collectives, the city council's heads of security and equality and a lawyer who represented the city council in its private prosecution efforts in the case. Interviews were also conducted with a number of experts on the issue: Montse Pineda, who is responsible for the Catalan approach to sexual violence; Laia Serra, a criminal lawyer specialising in human rights and gender; and, finally, Sara Ortiz, specialising in urban design from a feminist perspective. In addition to this fieldwork, we carried out an exhaustive content analysis of the judicial decisions issued in the case. More specifically, analyses were conducted of the first decision issued by the trial court, the Audiencia Nacional de Navarra, and of the second decision issued by Navarra's Tribunal Superior de Justicia. Finally, we have followed the developments in this and other cases of sexual violence, as well as the feminist movement's strategies.

Background: from sexual assault in the festive space of San Fermín to feminist resistance to sexist and sexual violence

The case that forms the basis for the analysis in this chapter occurred during the festival of San Fermín, a celebration of religious origin dedicated to the patron saint of the city of Iruñea. The event includes the famous running of the bulls, where a crowd made up predominantly of men run a distance of 849 metres, pursued by six bulls. This public performance serves to reinforce traditional notions of masculinity. Although the celebration has traditional religious roots, it now includes concerts, DJ sessions and other events that give it the character of a contemporary outdoor festival.

This outdoor public celebration takes over the streets of the city for nine days and nights. During the *fiestas*, Iruñea is transformed, as it hosts three times its usual population of 350,000 inhabitants (Roig, 2017). During the rest of the year, certain bars and nightclubs might cater to specific segments of the population, and they might be marked by certain styles of music or gender dynamics. The large crowds during the festival blur these distinctions (Roig, 2017), turning the city as a whole into one big unfettered celebration.

According to Velte (2019), for the nine days of the *fiestas* of Iruñea, the normal functioning of the city is suspended. Nonetheless, during the biggest social gathering of the year, the power relations that govern everyday society remain in force, whether they are based on gender, economic class or place of origin. In such a context, not everyone is able to enjoy the celebration in the same way, nor is everyone able to access the same spaces or engage in the same kinds of behaviour. At the same time, social and political conflicts tend to be reflected during the *fiestas*. In such a context, not everyone is able to enjoy the celebration in the same way or access all the spaces, even during the events. Sexist violence was an invisible conflict until 2013, when the issue finally came to the forefront of the municipal agenda.[5] That was the year that images of men groping women during the Txupinazo appeared in the press.

Meanwhile, celebrations like San Fermín tend to trigger certain mechanisms that reproduce patterns of violence against women. While the goal for some on a night of partying might be nothing more than fun and social interaction, others define a successful night out in terms of gaining sexual attention or affection. The flirting that occurs in these contexts is shaped by heteronormative patterns, and it often leads to harassment in the form of invasion of personal space, unwanted groping and the cornering or trapping of women. The celebration is founded on a model of heterosexism, transphobia and homophobia (Roig, 2017), creating an atmosphere that makes it difficult for some to enjoy the festive space. Although these *fiestas* are viewed as spaces for collective fun and enjoyment, not everyone is able to enjoy them under equal conditions.

The acts of sexual violence that occur during these events are examples of mechanisms that exclude women from public space and punish them for occupying spaces that, according to gender norms, do not 'belong' to them. Although the true scale of the problem is hidden because so many crimes go unreported, the data show that the situation is dire. According to information from the Spanish Interior Ministry, the total number of complaints registered between 2004 and 2017 was 74. The number of complaints registered by the police has been increasing significantly, especially since 2016, when the assault discussed in this chapter took place. Research has shown (Amnistia Internacional, 2018; Zuloaga et al., 2018) that despite this considerable increase in the number of complaints filed for crimes against women's sexual autonomy over the past few years, still only a small percentage of these aggressions are ever reported to the police. Additionally, these data are not always gathered accurately. Some studies (Zuloaga et al., 2018; Ministerio del Interior, 2018) do point to a slight but notable growth in the tendency to report such crimes in Iruñea. This cultural shift towards greater reporting of these crimes can be seen in Spain as a whole. Cases like the one analysed in this chapter seem to have helped raise awareness of the issue of sexual violence.[6]

All of this needs to be placed in the broader Spanish context, where sexual and other kinds of sexist violence represent a social problem that the public and the government have yet to take the necessary steps to eradicate. Public policy measures to address sexual violence in particular have been lacking. Evidence of this are the 8,200 sexual assaults with penetration that have been registered in Spain since 2009, a figure that works out to three such assaults per day (Ministerio del Interior, 2018). The refusal to make sexual violence a political priority means that each individual assault is seen as a unique event – an isolated occurrence (Barjola, 2018). This stance serves to de-politicise sexual violence, separating it from the structural context that it shares with other kinds of sexist violence. This failure to confront sexist violence might seem paradoxical in Spain, a country that was at the forefront of passing legislation against sexist violence (2004) and in favour of equal rights for men and women (2007). However, these legislative breakthroughs were ineffective, as the Spanish state has maintained its strong ties to the Catholic Church and is marked by an underlying conservative culture.

In the face of a largely conservative state apparatus, the long-standing activities of Spain's feminist movements have played a critical role. It is this very history that accounts for the contemporary rise of feminism around Spain. More recently, the economic crisis and the 15 million movement, born in 2011, have helped fuel the feminist movement in the country. In any case, in 2017, the worldwide feminist movement took historic steps that began a new cycle, shining an unprecedented international spotlight on the struggle against all forms of sexist violence (García, 2017). One of the landmark events in this expanded feminist drive was the international strike called on March 8, 2017, under a rallying cry coined by Argentine feminists, "*Ni una menos!*" (Not one less!). Since then, the feminist movement in Spain has come together to form the Comisión 8 de Marzo (March 8 Commission), an organisation that helped ensure the repetition of the strike the ensuing years of 2017–2019. The issue of sexist violence remained a core concern in these subsequent editions of the strike.[7] Meanwhile, the #MeToo campaign that swept the online world in the autumn of 2017 was another turning point and more evidence of the growing worldwide struggle against patriarchal violence.

This context explains why the case described in this chapter had such a large social impact. The feminist movement was able to communicate its conception of sexual violence to a wider public, providing society with a framework for interpreting these crimes. At the various demonstrations held in cities around Spain in 2017 and 2018, many of them prompted by the series of judicial decisions issued in the case, protesters chanted slogans such as, "*Tranquila, hermana aquí está tu manada*" (Relax, sister, your wolf pack is here), "*Yo sí te creo*" (I do believe you), "*La manada somos nosotras*" (We are the wolf pack), "*Basta ya con la justicia patriarcal*" (No more patriarchal justice), "*No es abuso, es violación*" (It's not abuse; it's rape). The initial trial court's decision also sparked the online campaign #Cuéntalo (Tell your story), with over 50,000 women from around the world recounting the sexual aggressions they had suffered.

While it is true that the case discussed here was a turning point in the reporting of sexual and sexist violence in Spain, feminists in the city of Iruñea had been

Fiestas, public space and rape culture 31

working since the '90s to raise awareness of sexual aggression in San Fermín and to denounce these attacks. Specifically, the citizens' group Plataforma contra la Violencia Sexista, founded in 1998, has played a key role in these efforts, along with other feminist groups, such as Emakume Internazionalistak, Gora Iruñea!, Bilgune Feminista, Farrukas and Andrea-Lunes lila and local cultural associations, such as the Federación de Peñas de Iruñea.

These groups have implemented strategies to appropriate or reclaim spaces where women had never been fully welcomed. For example, the Farrukas feminist collective led night marches starting in 2016. Perhaps even more significant was an action taken by 60 women who entered the square in front of city hall at the moment of the Txupinazo, carrying a sign that said "Autodefensa Feminista" (Feminist Self-Defence). It is worth emphasising that the feminist movement and the struggle against sexual assault are not easy in a conservative, Catholic city with ties to the Spanish right, and the city's social movements have worked to forge their own networks of connections and to build a strong feminist resistance (Cañete, 2014). Thus when the group sexual assault discussed in this chapter occurred, there was already a great deal of awareness and preparation amongst the population of Iruñea (Velte, 2019).

Case study: sexual violence as a mechanism of dispossession of public space and feminism as a path towards re-appropriation

The first verdict in the case was issued by the trial court, the Audiencia Provincial de Navarra, on April 26, 2018. The defence in the trial argued that the men had engaged in consensual sexual relations. The accuser said that the men had offered to walk her to the car where she was planning to sleep that night and that she ended up being pushed into the entryway of a building and sexually assaulted. The court found the defendants guilty of sexual abuse but not sexual assault.[8] One of the judges even voted to acquit the defendants on all charges, writing that the sex had taken place in an atmosphere of "revelry" and "glee" and that the victim had been sexually aroused. Both the public prosecutors and the private accusers maintained during and after the trial that the crime should have been considered sexual assault because it had included violence and intimidation.

The decision by the Audiencia Nacional de Navarra is revealing in that it shows that while Spanish courts focus on consent when they try sex crimes, they view this issue from a patriarchal perspective that helps maintain and reproduce rape culture (Brownmiller, 1976). This trial followed a pattern that is all too familiar in this type of case. The focus was on the behaviour of the victim and whether she had said no or had tried to get away, not on the behaviour of the attackers and whether they had perceived that the woman had given her consent. An analysis of the trial in this sense is of great interest because it allows us to underline the importance of the links between the feminist reading of festive space and the sexual violence that occurs within it.

The finding that this gang rape had been a crime of continuous sexual abuse with the offenders having availed themselves of a situation of superiority, aggravated by carnal knowledge and assimilated behaviours (Articles 181.3 and 181.4 of the Spanish Criminal Code), not an assault, meant that the court found that the victim may have consented to sex but determined that this consent was not valid. The decision did refer to circumstances that suggest a context of intimidation, including the number of attackers, their age, their physical descriptions and the characteristics of the scene of the attack, but this was not enough for the court to conclude that intimidation had occurred. The decision argued that the crime had been one of sexual abuse with the offenders availing themselves of a situation of clear superiority, but that because the woman had not resisted, the acts did not constitute sexual assault. The decision as a whole was shot through with one of the chief long-standing myths connected to rape: the belief that when a woman does not actively resist a sexual assault, it follows that the woman has consented to sexual relations. The judges found that violence and intimidation only exist to the extent that a victim offers resistance. This interpretation is not supported by the language of the Spanish Criminal Code (Faraldo & Acale, 2018).

It is clear, then, that the court based its findings about consent in this case not only on the behaviour of the woman at the time of the assault but also on her behaviour in public and the fact that she had agreed to leave this festive space in the company of her attackers. This led the court to conclude that she had failed to set the proper boundaries or to take steps to protect herself. Thus the social myths connected to rape, the norms that circumscribe women's behaviour in public spaces, were at the core of the logic that led the court to conclude that the victim had consented (although her consent was not valid under the circumstances) and that no intimidation had occurred.

Amongst the evidence the judges weighed in the case were the accuser's testimony and that of the defendants, as well as the forensic reports. In its analysis of this evidence, the court placed an emphasis on three aspects of the woman's behaviour on the night of the assault. First, the judges' decisions clearly attach a great deal importance on the moment when the young woman met her aggressors at the celebration and decided to leave with them. The court found that the woman and her eventual attackers may have had a conversation of a sexual nature at that time and that the woman may have expressed an intention to have group sex with them. Second, the judges paid special attention to the route the group took after leaving the festival together. Specifically, they speculated about whether the woman was aware of the situation and about why she had not tried to escape, seek out a street with more light and/or call for help. Third, the trial placed a great deal of emphasis on how the woman had entered the space where the rape took place. The judges' decisions reveal a concern with whether the woman had voluntarily agreed to enter the doorway or whether she resisted or had been forced. The focus on these moments of transition from the public space of the festival to the private space where the attack occurred allowed the court to construct its legal argument as to whether the woman had consented.

From the language of the judges' decisions, it is apparent that the court's concept of consent, and the basis for its findings in the case, take for granted the idea that men should have access to women's bodies. The definition of consent here is far removed from that of a profoundly intimate act whose truth lies deep within the self. Instead, here it is a social construct (Fraisse, 2007), characterised by masculine dominance and by a denial of the possibility that consent must be renewed or that it can be withdrawn. Thus any small gesture of complicity or closeness (a kiss, merely being in the same space with a group of men, etc.) is legally interpreted as proof of the consensual nature of any sexual activity that occurs in private space. The transition from public, festive space to private space, then, was deemed evidence of consent. The reading of consent was thoroughly mediated by social myths about rape, as is evident in the court's close examination of the woman's behaviour in public.

In fact, in her testimony at trial, the young woman provided explanations for her behaviour in this transitional space, indicating that she could never have imagined what was about to happen to her and that she had not consented to the assault. The transcript shows that she was asked during the trial about whether she had sought help when she was walking with the men, she answered,

> No, no no. You know, I was annoyed, but I didn't expect what happened. By that time, I didn't think it was a good idea to ask for help, and in any case the street wasn't busy enough for someone ... to call out to someone, I mean.
> (Audiencia Nacional de Navarra, p. 49)

When she was asked about having kissed one of the aggressors before being forced into the doorway, she had to explain that the kiss had nothing to do with consenting to entering the building. "I don't know. It was just a kiss, and right when I was kissing him another one of the boys said, 'Let's go, let's go', so it wasn't like I was kissing him for five minutes" (Audiencia Nacional de Navarra, p. 53). The narratives here reveal how gender socialisation shapes ideas about consent. A feminist analysis of the woman's testimony shows how the court's version of events, and indeed the testimony itself, relied on the social legitimisation and naturalisation of sexual violence, as well as the historical practice of blaming women who are victims of assault while granting impunity to perpetrators (Serra, 2018). Therefore, in order to better understand sexual violence, it is necessary to revisit the concept of consent, as the prevailing ideas of sexual consent remain artefacts of patriarchy (Toledo & Pineda Lorenzo, 2016).

In short, then, the transition from public space (the street) to private space (the doorway) was read as evidence of consent to what went on to occur in the private space. The case grants a stamp of judicial approval to the naturalisation of the kind of low-intensity sexual assault that occurs in the public space of the festival, while at the same time defining private space as a place of unrestricted male dominance. Men's behaviour in these spaces has nothing to do with sexuality. Instead, it responds to a need to maintain a social order characterised by male dominance, where men exert control over both physical and symbolic spaces. Thus, women are not only symbolically relegated to private space (Pérez-Orozco,

2014) but also dispossessed and expropriated of any claim to social space. This logic permeates the judicial apparatus and the legal arguments it employs, thus turning the courts into yet another mechanism to expropriate and dispossess women's lives.

The feminist movement has created a series of counter-narratives in response to women's dispossession from public space and the strategy of fear produced and reproduced by sexual violence in public and festive spaces. These feminist strategies to resignify fear are grounded in everyday life, and they are aimed at dismantling the apparatus of control and surveillance of women's bodies and attitudes (Barjola, 2018). In the case of San Fermín, these strategies have taken the form of the appropriation of public space through the placing of signs and posters, even in the most symbolically important locations for the festival, spaces that are also the most highly masculinised and thus the least welcoming of femininity. They occupy the space and draw attention to sexual violence in order to help prevent it and to undermine the arguments used to justify these attacks.

The traditional, dominant narrative on sexual violence functions to cement and entrench behaviours in women that give rise to a system of blame and victimhood, especially in the context of public celebrations. The recent boost in visibility of this phenomenon, however, has prompted many to identify as sexual violence behaviours that would have traditionally been accepted as natural, would have been seen as part of the price women have to pay merely because they are women and choose to attend such events. This new awareness has also changed the way women position themselves in response to this violence. This represents an obstacle to the principal function of rape and sexual violence, which is none other than to maintain women in a state of constant fear (Brownmiller, 1976). This is a way of deactivating sexual violence as a tool of social control to be used against women who usurp the culturally dominant prerogatives defined by those who exert dominance (Reynolds, 1974).

This increased awareness of sexual violence and the shift in how women position themselves in response are also due in large part to the work of the feminist movement. Feminists have taken the opportunity to teach the public about this issue as they have worked to refute the arguments in the court's decision and to criticise the judicial system as a whole. Far from demanding more jail time, the feminist movement has posed a challenge to the justice system's underlying patriarchal logic. It is true that court decisions have met with feminist resistance, but we cannot close this chapter without mentioning a third decision issued in this case by the Spanish Supreme Court, the country's highest court, which found that the crime was sexual assault, not abuse.

Conclusions

As this text has made clear, the case analysed here is a paradigmatic one for a number of reasons, chiefly because it has helped call attention to the need to draw connections between a feminist reading of space and festive contexts on the one hand and sexual violence on the other. The case has underlined the need for a new theoretical framework to rethink and question the dominant concept of consent and to shed light on how the judicial interpretation of the concept is tainted by a patriarchal reading.

This chapter has shown that when courts apply a patriarchal logic to their reading of the events that occur in public space, they feed into the rape culture that is prevalent at the festival of San Fermín and many other celebrations. However, this text has also made clear that the feminist movement's interpretation of the case has served not only to reclaim public spaces during the fiestas but also to refute the arguments that reinforce sexual terror (Barjola, 2018) and to help women reposition themselves in response to sexual violence. In other words, women no longer act as victims but as active agents working to reclaim these spaces.

It is clear that crowded outdoor festive spaces like that of the festival of San Fermín are far from gender-neutral. On the contrary, gender inequality tends to be accentuated in such spaces. While these events facilitate the exercise of highly normative masculinity without any social consequences, for women, they represent a loss of freedom and autonomy. The feminist movement's occupation of festive public space and the opportunities to intensify the resignification of concepts such as fear, sexual violence and consent are all fundamental steps towards a world in which women can enjoy celebrations and, indeed, life itself with greater freedom.

This chapter paves the way for future research on the ideas of appropriation and dispossession of space and on the patriarchal mechanisms at work in this process. It points towards a possible rethinking of the theory of dispossession in relation to women's lives (Federici, 2010; Harvey, 2003) – an approach that is not limited to economics or resources. This dispossession is also symbolic and related to access to public space and festive contexts. Thus it is critical to highlight the key role of sexual control and masculine domination (Bourdieu, 2000) in this web of relations.

Notes

1 Although the case became well-known under the name "La Manada" (the Wolf Pack), the authors will refer to the case in this text as a group sexual assault. The use of the expression 'Wolf Pack' reinforces the notion of a phantom aggressor. Additionally, this was the name used by the perpetrators themselves to signify their masculinity and dominance.
2 Called Pamplona in Spanish and located in Navarra, part of the greater Basque Country.
3 The Txupinazo has traditionally been a highly masculine space and has come to be a symbol of sexist assaults.
4 The authors define sexual violence as a structural phenomenon that ranges from verbal harassment to forced penetration, including various types of coercion ranging from social pressure to forced intimidation (Hopkins, 1984). Thus, according to our understanding of the term, attacks can constitute sexual violence not only when they affect a woman's body but also when they infringe upon her privacy. At the same time, we

define rape not only as penetration and use of force, as we understand rape to include any kind of use and abuse of a person's body (usually that of a girl or a woman) by another person (usually a man) without consent (Segato, 2016; Velázquez, 2003). Thus rape can have a number of different functions. It can be used to enhance masculinity (Segato, 2016) or to establish comradery amongst friends (Hopkins, 1984).

5 In 2015, an alliance of left-wing and pro-independence parties wrested control of the city council from the conservative Unión del Pueblo Navarro party, bringing about greater willingness on the part of the city to combat sexual assault.

6 It is important to note that the data collected by the police cannot be considered sufficient indicators of the multiple kinds of violence suffered by women. The events reported to the police are not necessarily representative of the cases that have occurred. We know that the real figures are much greater, but we cannot accurately count them (Zuloaga et al., 2018).

7 It is worth noting that the feminist movement in the greater Basque Country has always had its own organisation, autonomous and distinct from that of the rest of Spain.

8 Spanish criminal law contemplates crimes against the sexual freedom or integrity of adults, and it draws a distinction between sexual assault and sexual abuse. This distinction is artificial and does not take into account the continuum of sexual violence (Serra, 2018). Both crimes occur in the absence of consent from the victim. In the case of sexual assault, the lack of consent is implicit because the attack is carried out with the use of violence or intimidation. Sexual abuse is considered to be committed without violence or intimidation, but in the absence of consent. The law also contemplates the existence of sexual abuse when consent has been given but is invalid because the perpetrator "availed himself of a situation of clear superiority that restricts the victim's freedom" (art. 183.3 of the Spanish Criminal Code). The crime of which the attackers were convicted in the first two verdicts was this latter one. It is worth highlighting that the penalties for sexual assault and sexual abuse are very different (Faraldo & Acale, 2018).

Bibliography

Amnistia Internacional. (2018). *Ya es hora de que me creas. Un sistema que cuestiona y desprotege a las víctimas*. Madrid: Aministía Internacional España.

Barjola, N. (2018). *Microfísica sexista del poder. El caso Alcasser y la construcción del terror sexual*. Barcelona: Virus editorial.

Bourdieu, P. (2000). *La dominación masculina*. Barcelona: Anagrama.

Breger, M. L. (2014). Transforming cultural norms of sexual violence against women. *Journal of Research in Gender Studies*, 4(2), 39–51.

Brownmiller, S. (1976). *Against our will: Men, women and rape*. New York: Simon and Schuster.

Burt, M. R. (1980). Cultural myths and supports for rape. *Journal of Personality and Social Psychology*, 38, 217–230.

Cañete, S. (2014). Sanfermines: violencia machista y autodefensa feminista. *Pikara Magazine*, July 7. Retrieved from www.pikaramagazine.com/2014/07/sanfermines-violencia-machista-y-autodefensa-feminista/

Du Mont, J. D., Miller, K.-L., & Myhr, T. L. (2003). The role of "real rape" and "real victim" stereotypes in the police reporting *Practices of Sexually Assaulted Women. Violence Against Women*, 9(4), 466–486.

European Union Agency for Fundamental Rights. (2014). *Violence against women: An EU-wide survey*. Luxembourg: Publications Office of the European Union.

Falú, A. (2009). Violencias y discriminaciones en las ciudades. In A. Falú (Ed.), *Mujeres en la ciudad: De violencias y derechos* (pp. 15–38). Santiago de Chile: Red Mujer y Hábitat de América Latina. Ediciones SUR.

Faraldo, P., & Acale, M. (2018). *La Manada. Un antes y un después en la regulación de los delitos sexuales en España*. València: Tirant lo Blanch.

Farapi, S. L. (2009). *Análisis de las fiestas del territorio histórico de Gipuzkoa desde una perspectiva de género*. Donostia: Gipuzkoako Foru Aldundia.

Federici, S. (2010). *El calibán y la bruja. Mujeres, cuerpo y acumulación originaria*. Madrid: Traficantes de sueños.

Fraisse, G. (2007). *Du consentement*. Paris: Seuil.

García, J. (2017). 2017: resistencia feminista local y global frente a la guerra contra las mujeres y la austeridad. *Anuario de Movimientos Sociales 2015–2016*. Fundación Betiko. Retrieved from http://fundacionbetiko.org/wp-content/uploads/2018/02/03_grenzner.pdf

Griffin, S. (1979). *Rape: The power of consciousness*. New York: Harper and Row.

Haraway, D. (1998). Situated knowledges: The science question in feminism and the privilege of partial perspective. *Feminist Studies*, *14*(3), 575–599.

Harvey, D. (2003). *The new imperialism*. Oxford: University Press.

Hopkins, J. (1984). *Perspectives on rape and sexual assault*. New York: HarperCollins.

Hyams, M. (2003). Adolescent Latina bodyspaces: Making homegirls, home-bodies and homeplaces. *Antipode*, *35*, 535–558.

Isorna, M., & Rial, A. (2015). Drogas facilitadoras de asalto sexual y sumisión química. *Health and Addictions: Salud y drogas*, *15*(2), 137–150.

MacKinnon, C. (2018). *Feminismo inmodificado: Discursos sobre la vida y el derecho*. Argentina: Siglo veintiuno editores.

Ministerio del Interior. (2018). *Portal Estadístico de criminalidad. Indicadores de seguridad 2018*. Retrieved from https://estadisticasdecriminalidad.ses.mir.es

Patiño-Díe, M. (2016). La construcción social de los espacios del miedo: Prácticas e imaginarios de las mujeres en Lavapiés. *Documents d'Anàlisi Geogràfica*, *62*(2), 403–426.

Pérez-Orozco, A. (2014). *Subversión Feminista de la Economía: Aportes para un debate sobre el conflicto capital-vida*. Madrid: Traficantes de sueños.

Reynolds, J. (1974). Rape as social control. *Catalyst*, *8*, 62–67.

Roig, A. (dir). (2017). *Informe safermines*. Retrieved from www.pamplona.es/sites/default/files/2019-01/2017%20Noctambulas%20Pamplona%20doc%20final_AB.pdf

Rodó de Zárate, M., & Estivill i Castany, J. (2016). *¿La calle es mía? Poder, miedo y estrategias de empoderamiento de mujeres jóvenes en un espacio público hostil*. Vitoria-Gasteiz: Emakunde.

Ruddick, S. (1996). Constructing difference in public spaces: Race, class, and gender as interlocking systems. *Urban Geography*, *17*, 132–151.

Segato, R. (2003). *Las estructuras elementales de la violencia*. Buenos Aires: Universidad Nacional de Quilmes.

Segato, R. (2016). *La guerra contra las mujeres*. Madrid: Traficantes de sueños.

Serra, L. (2018). *Juicio a la justicia patriarcal Pikara magazine*, May 7. Retrieved from www.pikaramagazine.com/2018/05/juicio-a-la-justicia-patriarcal/

Toledo, P., & Pineda Lorenzo, M. (2016). *Marc conceptual sobre les violències sexuals en l'abordatge de les violències sexuals a Catalunya*. Barcelona: Generalitat de Catalunya. Retrieved from http://dones.gencat.cat/web/.content/03_ambits/violencia_masclista/Estudis_VM/Abordatge_VS/vmMarc-Conceptual-violencies-sexuals.pdf

Valle, T. del. (1997). *Andamios para una nueva ciudad: Lecturas desde la antropología*. Madrid: Cátedra.

Velázquez, S. (2003). *Violencias cotidianas, violencia de género: escuchar, comprender, ayudar*. Buenos Aires: Paidós Ibérica.
Velte, S. (2019). *Yo sí te creo. La cultura de la violación y el caso de Sanfermines*. Tafalla: Txalaparta.
Vigarello, G. (1999). *Historia de la violación (siglos XVI-XX)*. Madrid: Cátedra.
Zuloaga, L., Francés, P., Alemán, E., García, L., Tirapu, X., & Jabat, E. (2018). *Agresiones y abusos sexuales en Sanfermines. Estudio diagnóstico de las dimensiones y de los posicionamientos mediáticos e institucionales ante el problema*. Pamplona: Universidad Pública de Navarra.

Legislation

Convención sobre la eliminación de todas las formas de discriminación contra la mujer. Nueva York, 18 de diciembre de 1979.
Council of Europe. Council of Europe Convention on preventing and combating violence against women and domèstic violence (Istanbul Convention): 12 steps to comply. (En línia). Retrieved from https://rm.coe.int/CoERMPublicCommonSearchServices/DisplayDCTMContent?documentId=090000168046e809
Ley Orgánica 1/2004, de 28 de diciembre, de Medidas de Protección Integral contra la Violencia de Género (BOE, núm.313, 29.12.2004, pp. 42166 a 42197).
Ley Orgánica 3/2007, de 22 de marzo, para la igualdad efectiva de mujeres y hombres (BOE, núm. 71, de 23.03.2007, pp. 12611 a 126645).
Llei 4/2015, de 27 d'abril, de l'Estatut de la víctima del delicte. (BOE, núm. 101, 28.04.2015, pp. 36569 a 36598).
Sentencia de la Audiencia Nacional de Navarra (Sección 2), num. 000038/2018 de 20 de marzo de 2018.
Sentencia del Tribunal Superior de Justicia de Navarra (Sala de lo civil y lo penal), num.0000007/2018 de 4 de diciembre de 2018.

4 Sexual violence in Nigerian MusiCultural festivals
Narratives of the unheard victims

Richard A. Aborisade

Introduction

Sexual violence in Nigeria, like in most other parts of the world, is about the most under-reported crime in the country (Amaka-Okafor, 2013; Ezechi et al., 2016; Nwafor & Akhiwu, 2019; The World Bank, 2019). By implication, although there are no reliable statistics, sexual-related violence suffered by girls and women who attend various music and cultural festivals will continue to swell up the country's 'dark figures' of crimes and, specifically, sexual crimes. Meanwhile, literature on rape and other forms of sexual violence in the country identified social perception and attitudes towards rape as not only major barriers to the reporting of rape but also factors that fuel incidences of sexual violence (Aborisade, 2014a; Folayan et al., 2014; Africa Network for Environment & Economic Justice, 2016).

Social attitudes towards sexual violence in Nigerian societies, often characterised by blaming the victims, is believed to be linked to traditional gender-role stereotypes that are particularly related to sexual behaviours (Aborisade, 2014b; Nwafor & Akhiwu, 2019). For example, while men are expected to act out their sexual desire, women are not supposed to express their sexual interest, making their refusal to often be interpreted as a token. Furthermore, men are oriented to think that in certain circumstances, controlling their sexual urge is not necessary, such as dating situations and marriage (Amaka-Okafor, 2013). Consequently, there are several extremely negative implications that have emanated from the apparent tolerance towards rape, as victims resort to blaming themselves for the assault. This leaves women to be perpetually vulnerable to getting sexually harassed and assaulted at different scenarios, including social events and music-cultural festivals with little or no social support system for survivors.

Indeed, music and cultural festivals form a significant aspect of Nigerian social and cultural life. Such events offer opportunities for both young and old, men and women, Nigerians and non-Nigerians, at home or abroad to congregate to see a favourite musician, explore and perform aspects of their identity, cement their social relationships and, perhaps, engage in transgressive behaviour. Reports have shown that music and cultural festivals in the country portend risks and are sites of violence and harm (Endong, 2017; Premium Times, 2015). The high incidences of physical and sexual violence at music and cultural festivals in the country have been attributed to crowd dynamics, inadequate crowd monitoring, high volume of alcohol consumption and other factors. In particular, music and cultural festivals in Lagos, Calabar, Kaduna and Port-Harcourt have been identified as prone to incidences of gender-based violence, although there are no official statistics to support the avalanche of newspaper claims.

In this chapter, the accounts of survivors of sexual violence in various music and cultural festivals across the country are analysed. Hitherto, the apparent silence of women who suffer victimisation at these events creates the notion that such festivals are devoid of gender-related violence. Partly, this belief is premised on the notion that women who attend such music festivals are more accommodating to unsolicited comments about their bodies and touches from men during dances. However, newspaper reports have suggested that there are occurrences of sexual violence in and around music and cultural festivals across the country that leave victims in devastating consequences. Consequently, using sociological and criminological approaches, this chapter attempts to cover the gap in knowledge on sexual violence at music and cultural festivals in Nigeria.

Theoretical orientations: gender, sexual violence and the 'culture of silence' in Nigeria

In this chapter, a cross-disciplinary approach was made to engage theoretical perspectives from a number of traditions of study to analyse the issue of sexual violence of girls and women at music and cultural festivals in Nigeria. A bricolage of different theoretical perspectives is necessary for the profiling of sexual violence at social events in Nigeria because no single theory is sufficient to provide an all-encompassing explanation for the multidimensional issue. In addition, this current study is exploratory in nature, as sexual violence in Nigeria is highly under-reported, especially sexual victimisation suffered at music and cultural events.

Sexual violence against women has continued to take central positions in the work of feminists both in and out of criminology; hence, theoretical orientation in this work will be incomplete without considering the applicability of the feminists' perspective to sexual violence in Nigeria. From a radical feminist perspective, Brownmiller (1975) considered rape and threat of rape as a way to keep women in their place under patriarchy. The sexual objectification of women in Nigerian societies has been pointed out as a factor that fuels sexual harassment.

In relating the feminists' stance to Nigerian society, sexual violence against women can be attributed to the differential power relations within the society. The explanation for the motivation for sexual assault has been located around the complex interplay between existing social structures, conventional attitudes and socialisation (Africa Network for Environment & Economic Justice, 2016; Ezechi, et al., 2016). Amaka-Okafor (2013), identified differential gender socialisation of men and women in a patriarchal Nigerian society in which a girl child is socialised to become a wife and mother as a factor that builds a culture of silence amongst women in regard to domestic abuse and sexual victimisation. Furthermore, sexual socialisation for girls and women is manifested and reinforced through the school curricula, media, sex stereotyped expectations, gender-specific child rearing practices and role definitions, which are all firmly rooted in most Nigerian societies. In these socialisation processes, women are perceived as sexual objects whose function is to satisfy the needs of men so that in some situations, sexual coercion is considered normal and acceptable in-role behaviour. This form of social perception of gender and sexual roles that is highly tolerant of rape has been identified as a factor that makes the eradication of rape to be difficult, as it reduces the likelihood of reporting the crime. It also reduces the likelihood of police officers and judges conducting thorough investigations into such assaults.

In another angle to the pervasiveness of sexual violence in Nigerian society, the broken window theory of policing offers an explanation. The adoption of broken window theory is hinged on the persistent failure of the Nigerian criminal justice system to address the scourge of sexual violence within the society. According to Wilson and Kelling (1982, 33), the policing of minor offences might reduce more serious crime. They asserted that "if a window in a building is broken and left unrepaired, all of the windows will soon be broken". The apparent failure of the Nigerian criminal justice system to decisively address the problem of sexual harassment and assaults at domestic and neighbourhood levels is attributed to the growing rate of sexual violence in the country. Notably, the often reported reluctance of the police to conduct proper investigation on cases of sexual violence is one of the factors that militate against eradication of rape in the country (Amaka-Okafor, 2013; Ezechi et al., 2016). Also, the inflicting of secondary victimisation through the ridiculing of victims of rape and inadequacy of forensic materials to provide hard evidence of rape accounts for the reluctance of victims to report cases of sexual victimisation (Aborisade, 2014a). Consequently, assailants perceive that the law will not punish their actions due to weak instrumentations of the law as it relates to sexual offences, resulting in victims feeling more helpless and unsafe.

In spite of the overwhelming acceptance of broken window theory in criminological discourse, the mechanism with which it operates has been described as somewhat hazy (Sheley, 2018). As noted by Samaha (2012), the broken window is not a monolithic idea of causation but rather a collection of various potential relationships between appearance and disorder. There is yet to be a consensus on the efficacy of the theory, and there is, at best, weak empirical evidence to support the theory on its broadest form. However, the theory is considered appropriate in capturing the causation and prevalence of sexual harassment and assaults in the Nigerian society. In particular, it is able to underline the influence of social perception and stereotypes on policing rape as being responsible for the growth of sexual violence in the country. Furthermore, the deficiency of the Nigerian police to provide security for the entire society and by extension offer adequate security at such music and cultural festivals is

Methods

The research adopted a qualitative approach as a result of the unexplored, emotive and complex nature of being a victim of sexual violence in a conservative Nigerian society. The aim of the research is to capture viewpoints and breadth of experiences against commonality and dominant discourses, and since sexual victimisation experience is not freely discussed in public, the research could not make use of group settings. However, the design of the in-depth interviews ensured that there was sufficient structure in the guide to enable the addressing of research questions, while flexibility was also ensured to allow participants to make use of their thought processes.

Ethical consideration

Emphasis was laid on the voluntary nature of the study throughout the recruitment and interview stages of the research. Also, after the interview sessions, the researcher handed out to the participants an information sheet detailing local and national sexual abuse and counselling services. The participants were also requested to complete informed consent forms while the information about the study was shared with them at least 48 hours before interviews. Assurances on anonymity and confidentiality were given to them, and, as a result, only the researcher knew the identities of the participants in the study, and all data were securely held in accordance with Olabisi Onabanjo University regulations. Ethical approval for the study was also granted by the Research Ethics Committee of Olabisi Onabanjo University.

Sample participants

The community of interest of the study were women who had experienced sexual violence during their participation in music and cultural festivals in Nigeria. Since there is a dearth of research in this topic area, there was a need to capture information-rich cases as well as attempts to engage participants that might have a range of experiences, attitudes and beliefs. Consequently, the use of maximum variation sampling was considered essential, as it is an approach that turns the problems of heterogeneity between individual cases in small samples from ostensibly weak to strong (Creswell, 2013).

Eligibility for the study was set at age 16 or over, both during the time they experienced sexual violence at the festival and at the point of recruitment. The sexual violence they experienced had to fit the definition of sexual harassment, abuse and rape as contained in the Nigeria Criminal Code. This was screened for prior to the interviews according to the self-report of the participants. The indepth interviews were mainly conducted in English, although some of the participants were more comfortable speaking *pidgin* English and local dialects. Due to the relevance of first-hand experience in the purpose of this study, people who offered third-party accounts were ineligible to participate.

Procedures and data collection

In the recruitment process for the study, there was no natural group or organisation to approach for direct recruitment of participants; therefore, the bid to optimise diversity made the researcher utilise a varied approach over a three-month period, including posters on university campuses, community venues, web advertisements and social media. There were also the options of being interviewed over the phone or Skype that were made to prospective participants. Willing participants contacted the researcher, and, after their consents were obtained, they were requested to complete socio-demographic questionnaires, which were aimed at informing the analysis and contextualising the narratives. Also, it assisted in providing information about sample diversity for the guiding of recruitment strategies.

At the time of conducting the semi-structured interviews, the researcher made use of a topic guide that emanated from the findings of previous research conducted on a related subject of sexual violence (Aborisade & Vaughan, 2014). In a bid to keep to the principles of qualitative research, the initial topic guide evolved in order for insights gained in early interviews to inform subsequent ones (King & Horrocks, 2010). The author personally conducted all the interviews, which ranged from 30–60 minutes in length, and audio-recorded the process. Thereafter, the recorded notes were transcribed *verbatim*, and the anonymised transcripts were imported into NVivo 12, which is the software that is used to organise and manage qualitative data. The recruitment, interviews and analysis were done simultaneously until the point of saturation was reached: "a point of diminishing returns where increasing sample size no longer contributes new evidence" (Ritchie et al., 2003, 78). In all, 17 interviews were conducted.

Data analysis

Content analysis of the interviews was done using qualitative software program NVivo, version 12. According to Krippendorff (2012), content analysis involves the probing of content and themes of text to uncover both definitions contained in the text and those that emerged through the analysis. Therefore, thematic categories were derived from both the theoretical constructs and the data from which they emerged. Coding for theoretical themes was made in the first pass, which has to do with the global themes that I expected to find within the disclosure of gender-based violence and victimisation. In the second pass, I coded for themes that emerged from the content of the data. In the presentation of findings, thematic headings were used while exemplar quotes from the narratives of the respondents were used for the illustration and supporting of the findings. The brackets after each quote contain pseudonyms that respondents picked to refer to their identities, ages and venues of the festivals where they were assaulted, together with the year they experienced the assault.

Results

Sample description

Data for the study were collected between February and April 2019, during which 17 interviews were conducted: 11 face-to-face, 4 over the telephone and 2 using Skype. There were more than 17 experiences shared by the participants, as some of them had been victims of sexual violence at festivals more than once. The majority of the participants were students of higher institutions (11/17), most were residents of Lagos state and their age ranged from 18 to 35. There was variety in terms of the type of sexual violence suffered, the severity, the number of assailants and their reactions to the violence experienced.

Brief descriptions of music and cultural festivals where participants experienced victimisation

The 17 participants of the study volunteered accounts of sexual harassment and assaults experienced in six music and cultural festivals across the country.

One Lagos Fiesta (formerly Lagos Countdown) is a celebration of culture, music and arts of the people of Lagos state. It is held every December and culminates on December 31 with musical performances from top artists and jamborees that usher in the New Year at the Eko Atlantic City, Victoria Island.

Calabar Carnival and Festival is an annual event that is held in the City of Calabar, Cross Rivers State, and is widely regarded as Africa's largest street party. It is a cultural festival that showcases traditional culture and heritage through music, drama, costumes and other cultural creativities. Held every December, it draws millions of participants from within and outside the country.

Ojude-Oba cultural festival is usually held on the third day of the Eid el Kabir (Sallah) celebration in the City of Ijebu-Ode Ogun State. It is a historical festival that started in 1892 and welcomes hundreds of thousands of people annually. Though it started as a religious festival, it has metamorphosed into a music and cultural festival. The one-day event showcases the donning of colourful festive costumes from various peers and age groups to pay tribute to the king, after which the celebration continues with different forms of entertainment.

The *Carniriv* is an annual festival held in the City of Port-Harcourt, Rivers State. The name of the festival was coined from the words 'carnival' and 'rivers'. The seven-day carnival is held in December as part of festivities leading to the Christmas celebration. This affords a lot of travellers and tourists opportunities to attend the events. The carnival features different cultural and musical performances with top musical artists across the country.

Kaduna musical festival started in 2015, with the maiden edition held at the Ahmadu Bello Stadium, Kaduna. It was designed to be an annual musical festival, but the violence that plagued the first edition has put subsequent editions on hold. Its maiden edition was held for two days but ended in violence in which several people were injured as police fired tear gas and shot into the air.

Findings

In a bid to provide context for the themes that follow, it is imperative for the readers to understand the behaviours that offenders of sexual violence used towards their victims and acts that victims considered to constitute sexual violence. When the participants were asked about these, the researcher found that participants had limited knowledge of the concepts of sexual violence, sexual harassment and sexual assault. Those that experienced sexual violence from dating partners and acquaintances at the venue of the social events did not initially acknowledge it as abuse. Most of them perceive sexual violence as unwarranted and forceful sexual overtures from strangers at such events. However, after they were enlightened that non-consensual and forceful sexual advances from acquaintances and dating partners are also under the scope of sexual victimisation, they had more experiences to share with the study.

In the process of the interviews, participants were observed to be quite comfortable describing their experiences, and from a thematic analysis of the narratives, the ways in which they suffered sexual victimisation during music and cultural festivals across the country were apparent. The themes discussed in the report presented next are forms of sexual violence and severity, relationship with perpetrator(s), factors responsible for vulnerability and action taken after the harassment or assault.

Violent sexual victimisation at music and cultural festivals (types and severity)

Participants expressed that they were on the receiving end of various forms and degrees of sexual victimisation at different times when they attended musical and cultural events across the country. All the participants expressed that they were subjected to sexual harassment in the forms of unsolicited comments about parts of their bodies; slapping/touching of their buttocks; touching/pressing of their breasts; lifting up of their skirts or other clothing to reveal their panties and inner wears; men pressing the lower part of their bodies on theirs, mostly during dances; and unwarranted hugs with the intention to fondle their breasts. In the words of an attendee of Kaduna music festival,

> In the middle of the chaos, everybody started running for their dear lives as police fired bullets everywhere. I made an attempt to pass an exit that was jam-packed as many people also tried to use the same gate. Suddenly I felt a strong hand underneath my shirt, before I could understand what was

happening, the man lifted my brassiere and cupped my breast in his hand and pressed it hard. I tried to look back as I shouted but he pressed his body hard on mine and against the person in front of me so I couldn't turn my head. When he eventually released me, I was too shocked to call people's attention. Beside, everybody was concerned about how they will escape from the chaos.

(Sarah, 27, Kaduna music festival, 2015)

Aside from those that reported various forms of sexual harassment, there were those who suffered the sexual assault of rape. Nine of the participants recounted how they were raped in different areas of the events while one suffered sexual coercion on her way out of the venue. All those who were raped at an event stated that their assaults took place in secluded parts of the venues. Three of them were raped while they looked for a convenient place to relieve themselves, as there was either no provision of toilets or the available toilet was filled or too dirty for them to use. Meanwhile, two participants described how they were led to solitary areas of the events centres by their acquaintance and sexually assaulted by them. Amaka narrated how she ran into a supposed safe shelter when trouble broke out in the events she attended, only to be assaulted in her hiding place. In the case of Shola, she was drugged and raped during the event, while Angela narrated that her acquaintance took advantage of her state of drunkenness and assaulted her.

Adeola stated that she was led at gun point to a lonely place by her assaulters while she was looking for her boyfriend who "was taking too long to return after he excused himself to visit the toilet". She was eventually gang raped by a group of four men at a solitary place near the event venue. As for the final participant and the only one who was assaulted outside an event venue, she narrated how she was forced to offer sex in exchange for transportation from the event venue in the wee hours of the morning:

> I asked for a lift from the man because I saw that there was another man and three ladies in the car. So I thought that there were couples and I will be safe. However, it was surprising to me and apparently other occupants of the vehicle when he drove to a lonely road and parked somewhere. He announced to all occupants that he was no longer interested in conveying us to town except we surrender ourselves to have sex with him. I heard him tell the other man, who appears to be his friend, to walk away with his girlfriend. Then, the rest of the three of us were left to be begging for our honours. When it became clear that the guy meant business, the other two ladies succumbed and allowed him have sex with them. I was alone hoping that he would get tired and spared me but he still insisted and threatened to leave me alone on the lonely road. Out of fear for my safety, I allowed him have sex with me. I pleaded with him to even use a condom but he had none. I cried all through as I have never been so humiliated all my life.

(Maria, 29, Lagos Countdown – Now One Lagos Fiesta, 2014)

Relationship with perpetrators of sexual violence

Seven participants stated that they were sexually assaulted by total strangers. However, one of them, Nkoyo, said that she'd developed a 'friendship' with the person who sexually harassed her at the festival in Calabar. She narrated that the man bought her drinks and some other refreshments in the course of their stay together at the event. It was when he made his intentions known that he wanted her to follow him to his place and she declined that "he started to forcefully fondle my breast, touched my buttocks and told me blankly that he was only recovering some of the money he has spent on me that day".

On the other hand, two of those who were sexually assaulted at the events were victimised by their acquaintances. While Folake was raped by her boyfriend of six months, Lillian was assaulted by a 'childhood friend' whom she shared no romantic involvement with prior to the day of the assault. She said he pleaded with her after the rape and explained how he had been sexually attracted to her for years prior to the incidence. She related further,

> I have actually heard him a few times make sexual remarks about my body, but I shrugged them off as banters of a friend. While he even asked me to remove my dress at the event in a very dark lonely corner, I thought he was up to one of his jokes until he removed his belt and flogged me continuously ... (she broke into tears and ended the call – the interview was conducted over the phone).
>
> (Lillian, 27, Carniriv Festival Port-Harcourt, 2018)

Aside from these two, the remaining participants stated they were assaulted sexually by strangers. However, Adeola stated that she still has her suspicions that her assault, though perpetrated by strangers, must have been masterminded by someone known to her:

> The way and manner with which I was approached by the guy that pulled a gun on me and asked me to follow him was suspicious. He actually came for me among several ladies around me at that time. I thought I was about to be robbed, as it is a common occurrence in shows like that ... When I got to their hideout and found that there were more guys waiting, I knew I was in serious trouble. Also, they recorded the entire rape process on their phones and mocked me all the way. It was so cruel that I believe someone known to me must have put them up to that.
>
> (Adeola, 26, One Lagos Fiesta, 2017)

Factors responsible for vulnerability to sexual victimisation at festivals

The participants expressed factors that they perceived made them vulnerable to be sexually harassed and assaulted. Although some of them blamed themselves for not being vigilant and for their actions that made them vulnerable, they were also able to identify factors that made their harassment and assault inevitable. These included inadequate security personnel; ineffective crowd control, insufficient illumination of venues; lack of security gadgets, such as CCTV; inadequate commercial transportation from event venues to various destinations; social stereotypes about women who attend events as being loose and 'available'; sexual objectification of women; absence of male escorts; and poor performance of the Nigerian criminal justice system in punishing rape perpetrators.

Maria, who was assaulted on her way out of Lagos Countdown musical event, was emphatic about the "poor arrangement made by the organisers for people's transportation out of the venue". She expressed further,

> I had enough money on me to get a commercial vehicle to take me out of there but there was none in sight and I needed to leave the arena quickly, as there are always stories of how ladies that remained at such venues are usually assaulted. So, I jumped at the available opportunity to take me out of the venue, not knowing that I was inadvertently walking into the danger I was trying to avoid.
> (Maria, 29, Lagos Countdown, Now One Lagos Fiesta, 2014)

All the participants were unanimous in their condemnation of security measures put in place by organisers of the various festivals. They expressed that the shows are usually held in open spaces where no sitting arrangements are made because of their bid to maximise profits. Consequently, people are crowded together without any attempt to control the crowd. This, according to some participants, makes women in such crowded spaces highly vulnerable to sexual harassments. According to one of the participants,

> The belief of some unruly guys at such events is that a lady that attends without a guy to guard them is available to willing guys to sexually play with. Sometimes when they ask if you are with a guy and you answer "no", you will see them salivating and looking at you in suggestive manners.
> (Tacha, 27, Carniriv Festival Port-Harcourt, 2018)

This probably explained why some of the participants selected "absence of male escorts or partners at the events" as one of the factors that made them vulnerable to being sexually harassed and assaulted.

Action taken by survivors and post-assault experience

None of the participants in the study reported their victimisation to the police or sought justice in any form. Those who suffered harassment and assault from strangers indicated that they never considered reporting their offenders to the police because they were unknown to them. On the other hand, those who were raped by their acquaintances and dating partners stated that they did not report them because they "do not have a good case". In the words of Folake, who was raped by her boyfriend at a musical show during Ojude-Oba festival,

> Since he was my boyfriend at that time, reporting will mean exposing myself to ridicule because not only was I dating him, but I also went to that show with him. So, people will blame me for dating and following him around if I was not ready to have sex with him.
> (Folake, 27, Ojude-Oba festival, 2018)

Folake's sexual assault was done by her boyfriend inside his car, which was located in the parking lot of the Ojude-Oba event. He claimed he did it under the influence of alcohol. She further stated that her relationship ended on the account of the assault. Both Folake and Lilian, who was assaulted by an acquaintance, expressed that they have forgiven their assaulters. Participants who were sexually assaulted and harassed by strangers indicated that their refusal to report to the assaults to the police was borne out of the low confidence they have in the police. They noted that stories of women with similar experiences of sexual victimisation usually ended with justice not being served. Therefore, there was no basis for them to report to the police. Participants also cited social stigmatisation as a reason for not reporting.

Discussion

In this chapter, research was undertaken to explore the sexual victimisation experienced by women attendees of music and cultural festivals across Nigeria, the severity of their assaults and the sociocultural factors that inhibit reporting such incidences. The findings demonstrated that women attendees of festivals and carnivals in Nigeria are exposed to various forms of sexual harassment and assaults, which are overtly and covertly meted out on women at the events. This manner of harassment ranges from unsolicited comments about their body parts to forceful touching or pressing of sensitive parts of their bodies. In addition, women attendees of music and cultural festivals are subjected to sexual assaults by strangers, their acquaintances and dating partners, mostly in secluded parts of the festival venues.

Ezenagu and Olatunji (2014) and Onyima (2016) have noted the risks that are inherent in festival events across the country, which they described as sites of violence and harm. Newspapers have equally reported the high incidences of physical and sexual violence at music and cultural festivals in the country, which they attributed to crowd dynamics, inadequate crowd monitoring and high volume of alcohol consumption (Premium Times, 2015; Online Nigeria, 2017). However, until this present exposition, the accounts of those who have suffered sexual violence at such events have been largely unheard. This is primarily due to the inhibitions women who suffer sexual violence face in publicly recounting their experiences and seeking justice (Nwafor & Akhiwu, 2019). Literature have reported incidences of secondary victimisations that sexual assault victims faces in the hands of police officers across the country when they report their assault experiences (Aborisade, 2014a; Folayan, 2014; Ezechi et al., 2016). Also, rape victims are often blamed for making themselves vulnerable to being assaulted (Aborisade, 2014a), especially those who attend such festival events.

These findings are important in helping us to gain an understanding of why victims of sexual violence in music and cultural festivals in Nigeria do not report their victimisations to appropriate quarters. In relation to the feminist stance, sexual violence in Nigerian society is a product of the complex interplay between existing social structures, conventional attitudes and socialisation (Africa Network for Environment & Economic Justice, 2016). The differential gender socialisation of men and women in a patriarchal Nigerian society is also identified as a factor that builds a culture of silence amongst women to domestic abuse and sexual victimisation (Amaka-Okafor, 2013; Ezechi et al., 2016). Participants of the study believed that they would have been saved from assault if they'd had male partners as escort to the events. This belief of theirs may be connected to the socialisation lessons received by girl children that state that men are needed for protection, and a woman without a man is left vulnerable to societal scorn.

In alignment with the broken windows theory (Wilson & Kelling, 1982), the failure of the Nigerian police and the entire criminal justice system in the country to punish offenders of sexual crimes and protect victims from public ridicule has been found to be a key factor that instigates higher incidences of sexual violence and lower reporting rates by victims (Folayan et al., 2014; Africa Network for Environment & Economic Justice, 2016). Therefore, the perpetration of sexual violence at music and cultural festivals in Nigeria is a reflection of the pervasion of rape and sexual harassment in the entire society. This also partly explains the reluctance of the victims to report their assaults and harassment to the police, as people's confidence in the justice system to actually deliver justice and, thus, curb incidences of sexual crimes in the country is low.

The participants in this qualitative study were a much wider population than those generally targeted in bystander initiatives. Their experiences of sexual victimisation were in various regions across the country and at different music and cultural events. To genuinely develop a more effective response to sexual violence in Nigeria that recognises the complete context of the interplay between social structures, conventional attitudes and socialisation with regard to rape and silence, this broader population of victims needs to be engaged. A vital part of doing this is recognising the various manner of sexual harassment and assault used by offenders against women attendees of public events reported in this study. If better understanding of the secondary victimisation of this population is engendered, and the consequence of victims' silence to sexual violence is considered, an improved intervention for the amelioration of survivors of sexual crimes will be achieved.

Implications for theory, policy, practice and research

The findings of this research highlight the need for existing theories of sexual violence and assault to consider more explicitly the wider community context of music and cultural festivals within which perpetration and victimisation take place. There are indications from the findings of this research that, in addition to sexual crimes committed on the streets, neighbourhoods and other secluded areas, a whole range of gendered violence is perpetrated at social events and festivals across the country. While the feminist and broken windows theories are quite appropriate in discussing the social context and policy gaps in addressing the problem of sexual crime in Nigeria, the complexity of the culture of silence occasioned by secondary victimisation, social stigmatisation, gender stereotypes and lack of confidence in the justice system remains insufficiently represented. Therefore, there is a need for further development of existing theories to reflect this.

At present, there are no significant policies that support and protect survivors of sexual violence, especially against secondary victimisation and social stigmatisation. That may explain the reluctance of survivors to report their assaults to the justice system. This has served to preserve the culture of silence of sexual victimisation in the country. For police forces, in particular, better education of officers is needed to enlighten them about the importance of providing support for survivors of sexual crimes in the process of bringing perpetrators to justice. In addition, security measures at festivals have been found to be inadequate, making such events fertile ground for perpetrating crimes, especially gendered violence. Therefore, there is a need for the deployment of modern security apparatuses at festivals to prevent gendered violence in addition to other security needs.

Further research needs to be conducted to triangulate the perspectives of festival organisers and security personnel with that of survivors and perpetrators so that the problem of sexual harassment and assault at social events and festivals can considered from multiple viewpoints so as to broaden our understanding of the nature, prevalence, typologies, effects and disclosure of gendered violence at festivals.

Conclusion

Although previous research has highlighted the problem of sexual violence in Nigerian society and its inherent effects, rarely have women attendees of music and cultural festivals been directly studied within the scope of being victims of sexual violence. This research has captured rich experiences of women who were victimised during their attendance at music and cultural festivals across Nigeria. The findings indicate that there are a variety of ways by which perpetrators of sexual crimes subject women attendees to sexual harassment and assault. In spite of this, the victims are hindered from reporting and seeking help or justice because of the negative social perception of rape victims, secondary victimisation from police officers and social stigmatisation. The forms of sexual harassment and assault are similar to those survivors frequently report. However, survivors of sexual harassment and assault at music and cultural festivals report a higher rate of stranger rape than victims of sexual violence in other scenarios. Recognition of the predicament of victims of sexual violence at music and cultural festivals in Nigeria and the provision of social support would encourage reporting of sexual crimes in the country. This would promote an important double gain, reduce the 'dark figure' of sexual crimes and eradicate the 'culture of rape' in the country.

References

Aborisade, R. A. (2014a). Barriers to rape reporting for Nigerian women: The case of female university students. *International Journal of Criminology and Sociological Theory*, 7(2), 1–14.

Aborisade, R. A. (2014b). "It couldn't have been rape": How social perception and rape scripts influence unacknowledged sexual assault in Nigeria. *Research on Humanities and Social Sciences*, 4(8), 125–134.

Aborisade, R. A., & Vaughan, F. (2014). The victimology of rape in Nigeria: Examining victims' post-assault experiences and adjustment patterns. *African Journal for the Psychological Study of Social Issues*, 17(2), 140–155.

Africa Network for Environment & Economic Justice. (2016). *Why is there so much rape in Nigeria today?* ANEEJ Nigeria Retrieved July 22, 2019 http://www.aneej.org/much-rape-nigeria-today/

Amaka-Okafor, V. (2013). Nigeria has a rape culture too. *Guardian Africa Network*, January 14. Retrieved July 11, 2019, from www.theguardian.com/world/2013/jan/14/nigeria-rape-india-culture

Brownmiller, S. (1975). *Against our will: Men, women, and rape*. New York, NY: Simon & Schuster.

Creswell, J. (2013). *Data collection: Qualitative inquiry and research design: Choosing among five approaches* (3rd ed.). London: Sage Publications.

Endong, F. P. (2017). Nigerianess versus foreignness in the calabar festival and carnival calabar. *International Journal of English, Literature and Social Science*, 2(3), 4–15.

Ezechi, O. C., Musa, Z. A., David, A. N., Wapmuk, A. E., Gbajabiamila, T. A., Idigbe, I. E., & Ujah, I. A. (2016). Trends and patterns of sexual assaults in Lagos South-Western Nigeria. *The Pan African Medical Journal*, 24, 261. doi:10.11604/pamj.2016.24.261.9172

Ezenagu, N., & Olatunji, T. (2014). Harnessing Awka traditional festival for tourism promotion. *Global Journal of Arts Humanities and Social Sciences*, 2(5), 43–56.

Folayan, M., Odetoyinbo, M., Harrison, A., & Brown, B. (2014). Rape in Nigeria: A silent epidemic among adolescents with implications for HIV infection. *Global Health Action*. doi:10.3402/gha.v7.25583

King, N., & Horrocks, C. (2010). *Interviews in qualitative research*. London: Sage Publications.

Krippendorff, K. (2012). *Content analysis: An introduction to its methodology* (3rd ed.). Thousand Oaks, CA: Sage Publications.

Nwafor, C. C., & Akhiwu, W. O. (2019). Medicolegal analysis of sexual assault victims in Benin, Nigeria. *Nigerian Journal of Clinical Research*, 8(13), 10–17.

Online Nigeria. (2017). *Eyo festival ended in violence as masquerades fight each other in Lagos*. Retrieved June 28, 2019, from https://site.onlinenigeria.com/stories/206621-eyo-festival-ended-in-violence-as-maquerades-fight-each-other-in-lagos-video.html

Onyima, B. N. (2016). Nigerian cultural heritage: Preservation, challenges and prospects. *Ogirisi: A New Journal of African Studies*, 12, 273–292. http://dx.doi.org/10.4314/og.v12i1.15

Premium Times. (2015). *Kaduna music festival ends in violence*. Retrieved July 11, 2019, from www.premiumtimesng.com/news/top-news/195120-breaking-kaduna-music-festival-ends-in-violence.html

Ritchie, J., Lewis, J., & Elam, G. (2003). Designing and selecting samples. In J. Ritchie & J. Lewis (Eds.), *Qualitative research practice: A guide for social science students and researchers* (pp. 77–108). London: Sage Publications.

Samaha, A. (2012). Regulation for the sake of appearance. *Havard Law Review, 125*, 1563.

Sheley, E. (2018). A broken windows theory of sexual assault enforcement. *The Journal of Criminal Law and Criminology, 108*(3), 455–510.

Wilson, J. Q., & Kelling, G. K. (1982). Broken windows: The police and neighborhood safety. *Atlantic Monthly, 38*.

The World Bank. (2019). *Gender-based violence: An analysis of the implications for the Nigeria for women project*. Washington, DC: International Bank for Reconstruction and Development/The World Bank.

5 Between patriarchy and capitalism

The gendered violence of Intwasa International Arts Festival, Bulawayo, Zimbabwe

Khanyile Mlotshwa

Introduction

Playwright, scriptwriter and theatre director Ms Thoko Zulu's story of how she was hounded out of the Intwasa Arts Festival's director post partly sums up the story of women and women artists' relationship with Zimbabwe's second-biggest arts festival:

> Men feel challenged, creatively and intellectually, to deal with a woman in a position of power. That is why they got rid of me. They ganged up against me, and fabricated stories creating a media frenzy that I had embezzled festival funds. It was all meant to shoot me down. The fabricated stories came after I complained about what I felt was sexual harassment. A board member would call me every morning, asking about the colour of my panties. At one point when I was in Harare for the Harare International Festival of the Arts (HIFA), another phoned me around 2 am asking about the size of my bed. Male counterparts who were leading subsectors had no respect for me treating me like a dog. They would come to the festival office and insult me. I was like a leaper. There were even subtle witchcraft efforts to scare me. Every morning, and you can confirm this with the lady who was my secretary, as she is the one who alerted me to it; there would be some funny stuff on the door of the office. Next thing I heard that my contract was for only one year and that was it. My exit was staged. I suspect they were challenged by what I could have become had that opportunity not been taken away from me. They had to gang up, and shoot me down.
> (Thoko Zulu, interview with writer)

From Ms Zulu's story, it is clear that the gendered violence that women in Bulawayo – both as artists and as audiences – contend with is physical, psychological and epistemic. This chapter contribution discusses this gendered violence in the context of the Intwasa Festival of Arts KoBulawayo, an international festival second in size only to the HIFA in Zimbabwe. The festival ran for the first time in September 2005 when it was organised as a community project by artists in Bulawayo. Over the years, the festival has had to contend with the minimal participation of women in the festival, a development tied to the general arts industry in the city and the country. This underrepresentation of women in terms of participation in arts, in general, and the festival, in particular, is linked to the fact that women are oppressed in this society. Few families are keen on their girls participating in arts because such participation is seen as risky, including exposure to physical and sexual harm. In that most performances are at night, most people, especially parents, feel that it is not a good time for a young woman to be away from the safety of home. In the case of the Intwasa Arts Festival, its staging in venues mostly around the city, away from townships where most people live, has also been seen as excluding women in terms of accessibility of the performances of shows. The main shows of the festival are hosted in the city centre and mostly at night. Since a lot of people, including artists, live in townships or suburbs outside the city centre, this raises a challenge around transport home after the shows. The main shows of the festival have, therefore, become elitist spaces accessible to people who own cars, mostly men – including producers, directors, arts funders and male members of the audiences – who prey on the vulnerable young female artists.

The low numbers of women artists participating in the festival as alluded to by several interviewees in this chapter, who include the director and the administrator of the festival, are tied to the gendered epistemic violence of what it means to be a woman and to be an artist in Zimbabwe, in general, and in Bulawayo, in particular. As Thoko Zulu suggests, female artists in Bulawayo, and the larger Matabeleland provinces, face the tough task of self-definition as artists. In navigating the arts industry in the country, female artists are trying to negotiate a heavily patriarchal social system mostly run and maintained by men in terms of producers, directors, funders and benefactors. As Thoko Zulu points out in the later section of this chapter, the insistence is on merit, which is to say (female) gender and the disadvantage tied to it do not matter, women have to compete with men and prove that they are talented. On the face of it, that sounds fair enough. However, the challenge with meritocracy is that it underplays or overlooks the disadvantage against women, including the challenges they have convincing their families that they can be 'artists' and 'women'.

In all these struggles, they are trying to convince men. As Spivak (1988) points out, women are subalternised, meaning they cannot speak, and if they speak at all, they cannot be heard. Referring to women as the subaltern, Spivak theorises subalternity as a condition in which a social group cannot represent itself and is "removed from all lines of social mobility" such that it lacks agency (2005, 477). However, Spivak (2005) notes that subaltern does not just equate to "the oppressed" but specifically refers to the struggle to self-represent. This gendered epistemic injustice is what is at the heart of the participation of women in arts, in general, and the Intwasa Arts Festival, in particular. In Bulawayo, as part of Matabeleland, women are not only defined by men but also spoken for by men, meaning they are spoken *for* and spoken *of*. I have argued elsewhere that the category of women in Matabeleland is an effect of male representation (Mlotshwa, 2018). The Ndebele woman appears at the crossroads of the patriarchy of

Zimbabwean and Ndebele nationalisms as a "gendered and ethnicised collective subject position constructed at the intersection of social forces including history, geographic location" (Mlotshwa, 2018, 87–88). The woman artist in Bulawayo, as part of Matabeleland, is a gendered and ethnicised subject who appears already silenced by patriarchy, ethnicity and economic factors, amongst other issues. Her appearance on stage, on the page or the screen is after waging a struggle against all these violences.

This chapter begins by offering a background in terms of arts in Bulawayo and the Intwasa Arts Festival KoBulawayo. As part of this background, the chapter also discusses the situation of women in the arts in Bulawayo and Zimbabwe. The chapter then proceeds to give a theoretic perspective, discussing intersectionality, the coloniality of gender and epistemic injustice. The chapter uses in-depth interviews to collect data, and the interviews are analysed through a critical discourse analysis approach. In the last section of the chapter, I present the interviews and discuss them in the context of the literature discussed in the theoretical framework. The discussion of the interviews is divided into the current state of affairs, what is being done to stem the effects of the gendered violence of the staging of the festival and what would be the ideal situation.

Intwasa Arts Festival KoBulawayo and women in arts in Zimbabwe

According to its director, Mr Raisedon Baya, Intwasa Arts Festival KoBulawayo started off in September 2005 as a "loose coalition of the willing, community based artists wanting to celebrate the arts but evolved into one of the strongest arts and culture organization in Zimbabwe" (2016, 40). Intwasa is an isiNdebele word meaning 'spring'. The festival was created by artists and arts administrators in Bulawayo with the aim of celebrating creativity and diversity and empower women and young people through the arts (Baya, 2016, 40). Furthermore, the festival was meant to provide new markets and collaborative opportunities while marketing the City of Bulawayo as a tourist destination (Baya, 2016, 40). According to the organisers, over the years, the festival has hosted several international artists, hosted international workshop facilitators and published literary works under the banner of the festival. The festival mostly runs for five days, covering an average of 45 arts events across genres in five different venues around Bulawayo. The venues are mostly dotted around the city centre, including the city hall, the city hall car park, the National Arts Gallery building, the Bulawayo Theatre and the Eveline High School hall. The festival covers genres that include music, literary arts, visual arts, dance, spoken word and theatre to film. In what sounds ideal, Baya notes that the festival gives opportunities to everyone, "from women to men, old and young, local and foreign artistes to come together and celebrate cultural diversity and human creativity" (2016, 40). This, however, underplays ways in which the arts industry, in general, and the Intwasa Arts Festival, in particular, excludes women, even in ways that it may not consider discriminatory. For example, when festival organisers come up with a programme like *Wine,*

Words and Women, what might not be obvious is that conditions are such that women cannot compete with men, and, therefore, they need their own space.

The participation of women in arts has always been shrouded in controversies that are epistemic, such as underrepresentation, symbolic annihilation and emotional and physical, such as sexual abuse and violence (Kamara, 2014; Bretthauer et al., 2007; Chari, 2008). In the case of theatre in Zimbabwe, Chivandikwa, Mhako-Mutonhodza and Sambo note that the craft is still "fraught with gender biases, contradictions and inconsistencies" (2010, 181). These challenges persist even though work on theatre has continuously and continually engaged the theme of gender in varied ways. Sibanda notes that, even though gender mainstreaming has made strides in Africa, "the creative industries have faced a myriad of challenges, including lack of institutional support and gender policies, gender-friendly spaces and cultural patriarchal values" (2018, 69). What the arts have tried to do is to capacitate women and support programmes for men, these have failed to address the imbalances that pervade the sector (Sibanda, 2018, 69). In terms of theatre, the focus has always been on issues of gender insensitivity, biases, stereotypes and the limited depth in character construction in the country's theatre (Chivandikwa et al., 2010, 181). The challenge for arts, in general, and theatre, in particular, is that most of the work on gender is occasional commissioned work (Chivandikwa et al., 2010, 181). However, it is important to emphasise that art producers "have a heavy responsibility in crafting dramatic themes and characters whose gender implications have a significant cultural and socializing function" (Chivandikwa et al., 2010, 182). Theatre production companies, such as Amakhosi Theatre, have engaged in gender mainstreaming efforts through various training programmes and performances (Sibanda, 2018, 69). However, results out of such efforts are not yet as visible since the industry is still characterised by inequalities between female and male artists. The persistence of these inequalities is visible, especially in representations in arts. Chivandikwa, Mhako-Mutonhodza and Sambo note that some artists have "consciously or unconsciously crafted images of femininity which are problematic in terms of their potential to subordinate or perpetuate the subordination of women to negative structures of patriarchy" (2010, 187). Zimbabwean theatre arts are seen as dominated by stereotypes of "prostitutes, frail and weak women, gossiping or noisy old women, materialistic wives, tyrannical female bosses or employers and ungrateful or unreasonable mothers-in law" (Chivandikwa et al., 2010, 187). For Chivandikwa, the patriarchy of theatre in Zimbabwe is clearly seen in that both creators and audiences engage in what can be characterised as the female gaze (2009, 183). What this raises is the challenge for playwrights, directors, designers, actors and spectators "to refashion femininities in theatrical discourses" (Chivandikwa et al., 2010, 189).

Intersectionality, coloniality of gender and epistemic injustice

Mendoza argues that postcolonial and decolonial feminisms – as anticolonial feminism – politically challenge "imperialist and colonizing practices, past and present" (2016, 100). This theoretical tradition emerges out of postcolonial and decolonial theories that are anti-colonial perspectives that critique modernity, colonialism, Eurocentrism, capitalism, nationalism and racism (Mendoza, 2016, 100). This theoretical tradition is traced in the works of Gayatri Spivak, Maria Lugones, Silvia Rivera Cusicanqui and Rita Segato, amongst others, who proffer an "alternative account of modernity as a violent process intricately tied to the construction and imposition of race and gender hierarchies" (Mendoza, 2016, 100). Both theories draw attention to the historical entanglements of race and capitalism (Mendoza, 2016, 101). Postcolonial and decolonial feminism brings in the gender perspective into this intersectional analysis. While postcolonial theory has developed as a men's-only club such that Spivak is seen as gate crashing, decolonial theory, in its development around the modernity/coloniality group, has embraced feminist scholarship integrating the work of feminists of colour (Mendoza, 2016, 101).

Further to this, Smith insists, "There is no black feminism without intersectionality" (1998, xxiii). The concept of intersectionality as developed by black feminists has been appreciated for redefining theoretical, political and methodological approaches to feminist research, and also criticised for being merely descriptive (Alexander-Floyd, 2012, 5). It was coined and developed by Kimberle Crenshaw and focussed on analysing the multi-dimensionality of women's oppression as opposed to the single-axis approach that has always been used. Cooper noted that intersectionality emerged in the 1980s as an "analytic frame" attending to the particular positionality of black women in civil rights law and within civil rights movements (2016, 385). Despite this, Mendoza notes that critics have argued that intersectionality adds nothing new to feminist literature except to replicate "long-standing problems of identity politics, overemphasizing decontextualized categories of identity, focusing too narrowly on a small subset of structural constraints, or overemphasizing racism within feminism" (2016, 104). Tomlinson has defended intersectionality from criticism that it is overly focussed on identity pointing out that "diminishing the role of power in identity formation, such critics demonstrate a desire for individual self-invention, as if history and power no longer have claims on us, as if the significance of identities lies in expressions of subjectivity" (2013, 1000). Some of the critics have been preoccupied with 'correcting' intersectionality's emphasis on black women or women of colour by extending its analysis to the study of all women and all people (Garry, 2011). However, intersectionality has resisted this 'conceptual inflation' to study multiple vectors of power, including whiteness, class and religion, arguing that it displaces and erases black women, returning white women to the centre of feminist theory (Mendoza, 2016, 104). It has been argued that postcolonial and decolonial feminism have embraced intersectionality such that they appear as intersectional, simultaneously addressing gender and race (Mendoza, 2016, 104). However, Mohanty et al. (1991) noted that intersectionality has been central to third world feminism that sought to confront colonialism. McClintock (1995) has argued that it is near impossible to understand colonialism and imperialism outside of the understanding of the invention of race.

Intersectionality is tied to issues of epistemic injustice through the observations that emerge out of the standpoint conceptual framework. Feminist thinkers came up with standpoint empiricism to address issues of bias in knowledge production. Harding differentiated between 'standpoint feminism' and 'feminist empiricism', arguing that 'feminist empiricism' sees male bias as 'bad science' that can be cured through rigour (Harding, 1986, 1991, 111–120). Intemann noted that, standpoint theory maintains that "biases could occur even in cases where traditional scientific methodologies were followed" (2016, 216). In that sense, bias is then a result of power structures and how they shape or limit the production of knowledge (Inteman, 2016, 216). This partly accounts for epistemic injustice. For Tuana, early work in standpoint theory focussed on "links between privilege/power and the nature and limits of knowledge" (2017, 125). It was informed by Marxist analyses addressing "the inextricable connections between politics and knowledge production" (Tuana, 2017, 125). For Fricker, epistemic injustice borders on "distributive unfairness in respect of epistemic goods such as information or education" and the question of whether all the people receive these epistemic goods equally (Fricker, 2007, 1). Fricker discussed two forms of epistemic injustice: testimonial injustice and hermeneutical injustice (Fricker, 2007, 1). Testimonial injustice is when "prejudice causes a hearer to give a deflated level of credibility to a speaker's word" (Fricker, 2007, 1). Hermeneutical injustice is when, at a prior stage, "a gap in collective interpretive resources puts someone at an unfair disadvantage when it comes to making sense of their social experiences" (Fricker, 2007, 1). To Pohlhaus, epistemic injustice considers how "epistemic practices and institutions may be deployed and structured in ways that are simultaneously infelicitous toward certain epistemic values [...] and unjust with regard to particular knowers" (2017, 13). Pohlhaus argued that feminist, critical race and decolonial scholars have for some time now been interrogating the "ethics and politics of knowledge practices" as part of epistemic resistance "through identifying and calling attention to ways in which knowers have been wronged in their capacities as knowers" (2017, 13). Issues of epistemic injustice are also central to feminist epistemology. In the standpoint approach, it was argued that "feminist theorizing be grounded in women's material activity and must be part of the political struggle necessary to develop areas of social life modelled on this activity" (Hartsock, 1983, 304). The aim here is to "transform the subject of knowledge in the sense of focusing on knowledge obscured by dominant interests and values" so as to undermine knowledge and practices implicated in oppression (Tuana, 2017, 125).

In-depth interviews and critical discourse analysis

This research is mainly based on in-depth interviews. The interview has been described as a dialogue and "an inter change of views between two persons conversing about a theme of mutual interest" (Kvale, 1996, 2). It is a "construction site of knowledge" (Kvale, 1996, 2). For this research, I conducted six in-depth interviews with various key informants regarding the Intwasa Arts Festival. I spoke to the first and former director of the festival, Ms Thoko Zulu, who is now based in South Africa where she is trying to build an arts career from the ground up. Ms Zulu is a scriptwriter, playwright and stage director. I spoke to the current director, Mr Raisedon Baya, because he has information about the festival. having run it since 2007. Mr Baya is a scriptwriter, playwright and director for both stage and screen. I also spoke to the administrator of the festival, Cynthia Runyararo Mutandi, who has been with the festival since its inception. She has a slightly longer institutional memory of the festival than the director. Finally, I spoke anonymously to three female artists who have participated in the festival. The artists represent different arts genres and different generations in terms of arts in Bulawayo. One of them is a playwright and filmmaker. The second one is an administrator and did acting and dancing some years back when she was younger. The third one is involved in playwriting, play producing and poetry (spoken word). The point was "to understand the world from the subject's points of view, to unfold the meaning of people's experiences" (Kvale, 1996, 1). The interviews were analysed through the critical discourse analysis (CDA) method in which they were closely read for emerging discourses on the Intwasa Arts Festival. CDA combines theoretical observations on the social nature of language, its function in contemporary society and close textual analysis as a form of social and cultural analysis mapping circulations of meaning (Fairclough, 1995, 53–54; Krijnen & Bauwel, 2015, 41). The aim here has been to trace the arch of patriarchy and gender violence against female artists at the Intwasa Arts Festival.

The gendered epistemic injustice of the Intwasa Arts Festival

As has been outlined in the introduction, the gendered violence at the Intwasa Arts Festival manifests itself in various ways. It includes exposure to possible sexual abuse. In that the most important and attractive shows that take place at the centre stage of the festival are mostly programmed for late at night, it exposes people to physical bodily harm as they try to find their way back home unless alternative secure transport is secured. For female artists, this might mean getting transport from producers, directors, arts funders or even members of the audience, thus exposing themselves to potential sexual harassment. For females in the townships who could be potential audience members, this simply means that they may not get to see the shows. This programming of shows, not only disadvantage women artists, who, in any case, culture and customs expect to be home at sunset like women brought up in 'proper homes', but also from the start close out budding artists. For many families, 'decent' girls should consider looking for a nine-to-five job. In a sense, for most women who might have a talent or may feel the stage is their space, their art careers are dead before they even start. This general climate of patriarchy and sexism, therefore, creates an atmosphere of gendered epistemic injustice for many female artists. We shall now discuss the interviews and the emerging discourses on the gendered violence of the Intwasa Arts Festival.

The current state of affairs

The first and former director of the festival, Thoko Zulu, believes there is rampant patriarchy and sexism in the arts, in general, and in the festival, in particular. She said this patriarchy and sexism is meant to crush women artists and ensure that they have no significant voice or input into what happens at the festival. Coming from someone who has been at the highest echelons of the festival and yet was made to feel powerless by males, some of whom were junior to her, her comments paint a vivid picture of the silencing that women contend with at the festival. Ms Zulu said that when she was the director of the festival, she realised that many artists, most of them men, felt challenged "creatively and intellectually" when dealing with a woman in a position of power. Her observation raises an important point: that it is not only men who are frightened or are suspicious of women as leaders in arts and creative spaces, such as Intwasa Arts Festival. This does not fall into the trap of the often bandied around stereotype that women do not support each other, which is why they are always led by men. However, it does point to the discursive and structuring power of patriarchy. Women who are suspicious or not supportive of female leadership are not this way because there is something wrong with female leaders as humans but because they have been conditioned to believe that the only credible leadership is male leadership. In a sense, patriarchy as a discourse in the arts is such that only men can write, direct and produce productions successfully because they are born leaders. Women have to struggle against many challenges to reach these positions. This is why the appearance of Ms Zulu at the helm of the Intwasa Arts Festival was not palpable amongst a number of artists, including females.

Upon reflection of how she left the festival, Ms Zulu said, "They gang up and either intimidate you and if you stand firm they shoot you down". Here she is referring to the hegemonic masculinity representing men as born leaders and therefore entitled to leadership. The other aspect of this notion is that even when they are not in leadership, men must not listen to a female leader because to start with, that woman is in the wrong place. Hegemonic masculinity is based on the "structural fact, the global dominance of men over women" (Connell, 1987, 183). Hegemonic masculinity is the practice that subordinates other forms of masculinity to maintain the patriarchal social order whose final implication is to facilitate the dominance of men over women (Connell, 1987, 183; Connell & Messerschmidt, 2005, 832). At the helm of the Intwasa Arts Festival, Ms Zulu had to contend with hegemonic masculinity where "the currently most honoured way of being a man" for most artists was to resist female leadership (Connell & Messerschmidt, 2005, 832).

The director, Mr Raisedon Baya, and the administrator of the festival, Miss Runyararo Cynthia Mutandi, speak about how work by women, consequently female artists, is scarce, and as a result they, struggle to balance the festival content. Miss Mutandi says there has always been a low turnout in terms of female artists' participation at Intwasa "despite the fact that the festival creates a conducive environment for female artists to showcase their work". Miss Mutandi said, as organisers, they put out calls for participation, organised competitions, workshops, exhibitions and registration for performance at the festival, but "the number of female artists who register for participation is generally low despite" (Runyararo Cynthia Mutandi, interview with the writer). As a result, the festival organisers have been forced to headhunt to improve the involvement of female artists in their programmes. What this brings up is how patriarchy operates as discussed earlier in this chapter, that it structures society in such a way that even though women are the majority in any society, they become a minority in participating in public spaces, such as the arts and the festival. This is tied to hegemonic masculinity, where men are given a sense of entitlement, which means they can access the public space without the hindrances that women have to contend with. Miss Mutandi said even though this strategy has improved the participation of women in the festival, there was still a need for improvement.

Ms Mutandi, however, did not feel that programming, the scheduling of the shows and the venues were part of what needs to be looked into and changed to achieve improvement in terms of women's participation in the festival. Miss Mutandi said the shows and the venues are accessible to all people, and the shows are slotted for times that are safe, meaning that "all venues are 'safe spaces', they are all inclusive, accessible and do not close out anyone, either as performers or member of the audience" (Runyararo Cynthia Mutandi, in an interview with the writer). However, it is not as simple and ideal as Miss Mutandi wants to argue. She further problematises the issue of programming, scheduling and venues, pointing out that to make the festival accessible to ordinary women in townships, most of the shows are put up "during the day or in the evening so as to accommodate as many people as possible". Realising that this is not enough as an explanation for challenges related the festival and women, Miss Mutandi further noted that economic challenges have disempowered female artists and female members of the audience who have no disposable income to participate in the festival.

This makes a lot of sense in the context of the availability of venues and performance arts venues close to where these women live, considering that nearly every township has a youth centre, a women's centre and a public hall where these shows could be mounted. It would be important for the festival to go out to places where a large number of women live and ensure that they also get to access the festival. Miss Mutandi's observation is also important in that when she talked about resource challenges that the festival faces in taking its shows to where the people live, she touched on the issue of the political economy of the festival. Broadly speaking, political economy is all about the "ways in which a society provides for its needs, including the need for art as cultural expression" (Bechtold et al., 2004, 461). Bechtold, Gunn and Hozic noted that the political economy of art is tied to "its role in creating culture" (2004, 461). In a sense, art "production, consumption, aesthetics and its social significance, public funding, and legislation all contribute to society's cultural reproduction" (Bechtold et al., 2004, 461). In this case, where festivals are all about the distribution and consumption of art, poverty and the perilous state of the country's economy incapacitates the Intwasa Arts Festival from ensuring that the cultural goods showcased at the festival can be consumed by all people equally. In that the face of poverty is female, this imbalance in accessing the cultural goods of the Intwasa Arts Festival is then gendered. In that the festival, as some kind of a public sphere, represents, which means it effectively 'fabricates' or 'creates' its own public, women are effectively closed out of this "cultural reproduction" (Bechtold et al., 2004, 461). As a social space, the arts are central to the construction of identities and belonging (Cottle, 2000, 2). I use representation in the constructionist sense, as explained in the next paragraph. This is important in the context of artistic themes dealt with in the festival shows.

In terms of artistic content, Miss Mutandi said an effort is made to get artistic productions that are 'gender sensitive' and 'inclusive', even "though more can still be done". She noted that the festival organisers have been working hard to think about how they can improve the presence of women in the festival in terms of content. In what borders on epistemic injustice or epistemic violence, the presence of women is thought of in two ways: first, it is about their physical presence at the festival, both as artists and as audience members; and second, it is about how they feature in artistic content at the festival. Miss Mutandi feels the challenges are industry-wise as "the festival happens once a year, and that platform alone is not enough, we can only do so much". The issue of the presence of women at the Intwasa Arts Festival is an issue of representation. Representation has been taken to mean to make present that which is not there. In a constructionist approach, Hall (1997) argues that representation is not simply about reflection, but platforms of representation are involved in 'constructing' whatever they represent. Webb noted that "the processes of representation do not simply make connections, relationships and identities visible: they actually make those connections, relationships and identities" (2009, 10). In considering seriously the question of female representation at the Intwasa Arts Festival, the questions border on ways in which the festival, as tied to the broader cultural reproduction of society, has implication for the subjectivity of women as cultural citizens. This is important in that representation is about how people are seen, treated and brought into the economy and about social power (Karppinen, 2008, 27–28). Ms Zulu noted that, besides symbolic annihilation on the stage, off stage, women artists are not spoken about or given the same prominence as their male counterparts.

Questions about safety and sexual harassment are some of the discourses that emerged in conversation with the organisers of the festival and the artists. Ms Zulu noted that the safety of the festival would always depend on the artists involved as a collective such that this boils down to "individuals and their understanding of their contractual rights and their experience in the industry and their 'back up' system". This is a very technical approach to the issue of safety. The current director of the festival, Mr Baya, proffered a solution, which he, however, believes should start from the grassroots of the industry where women should occupy positions to influence policies regarding the safety of fellow women. Mr Baya admitted that "sexual harassment is rampant in the sector. What needs to be done is to find safe places for victims to come out and talk". Mr Baya said there is also a need for "safe places" for female artists to work and produce their art "without harassment". However, for Baya, what could address these challenges in a huge way is to put women into positions where they are empowered as directors or producers. However, Mr Baya could not speak about the presence of sexual harassment within the context of the festival. There are several reasons for Baya's silence on this matter. It could be that there is a general silence in regard to this issue such that victims never report harassment, and administrators, therefore, lack any data or knowledge to deal with it. The second reason could be that sexual violence at the festivals, like in the rest of the industry, has been normalised such that it is hard to begin to speak about it, let alone tackle it. It is only strong women like Ms Zulu who can name it. This calls for an investment, in terms of policies, by Intwasa to create a female-friendly environment.

Ms Zulu said her approach to other female artists has always been merit over affirmative considerations around gender. She said she encourages other women not to play the 'gender card' and hold up "our gender as an entitlement to opportunities" (Thoko Zulu, interview with writer). The challenge with meritocracy is that it is not a solution to the challenges that women face in that it ignores the disadvantages that women put up with making them unable to compete with males. In a sense, at face value, this view by Ms Zulu would be interpreted as falling into the trap of patriarchy. Ms Zulu, however, acknowledges that meritocracy has been globally used to undermine women in the arts. She said men are seen as tackling global themes while "women usually choose or focus on petty and personal issues when writing. True? Not always" (Thoko Zulu, interview with the writer). Meritocracy is always upheld where women's work is concerned as opposed to men or male artists who are given an opportunity to try their work. In reference to the Intwasa Arts Festival, Ms Zulu, noted that the festival was created by artists, who, at the time, all lacked experience in organising and running festivals, such that any insistence on meritocracy would be "all male (tendencies) because 'their' plans have always been to 'make' the festival all male".

What is being done?

Both art administrators and artists at the Intwasa Arts Festival point to some efforts being made to address the challenges of gendered violence in relation to the festival. All these efforts are said to be focussed on making the festival female friendly. Miss Mutandi said over the years, the festival has recognised the inert epistemic injustice in the arts industry and in its own organisation. As a result, the Intwasa Arts Festival has implemented strategies to address the exclusion of women in the festival, including events such as Women, Wine and Words, which is a platform that focusses mainly on performances by female artists and workshops where female artists discuss their challenges. Miss Mutandi noted that workshops have made the festival organisers to realise that there were genuine challenges around women's "access and participation [...] at the festival". For the festival edition starting in September 2019, the organisers have introduced the Intwasa Women's Day, which is dedicated to activities for and by women in the arts. For Mr Baya, the paucity of content by female artists means that the festival organisers have to come up with more strategies to encourage the production of content by women artists.

Such programmes, which are organised as exclusively for women, usually face a number of challenges. The first challenge is funding. On the one hand, the programmes may find funding from non-governmental organisations that are always pushing seemingly politically correct programmes on female empowerment. However, such funding comes with its own challenges, which may lead to a further symbolic annihilation of women. The easiest example that comes to mind here involves issues of political economy in that, at times, funding is tied to specific themes that may not be the urgent issue to the women who receive such funding. Alternative sources of funding for the arts is from the corporate world, which, under conditions of an economic meltdown, has been reluctant to fund the arts, in general, and to fund women arts, in particular. The second challenge is that creating what looks like affirmative spaces for women's art can have the implication of reducing women's art to some form of inferior art that cannot compete in the mainstream. The solution to this can be insisting on a 50/50 split in gender in terms of programming within the mainstream sector of the festival.

The ideal scenario

The administrators and the artists also painted a compelling vision of what they would like the festival to be in terms of equality in the participation of males and females. Artists do not mince their words and insist that equality must start with the administration and the running of the festival. The theatre and film artist interviewed anonymously said she felt that there was a need to ensure that women also play a role in the compilation of the festival's programme through chairing some sub-committees. It is mostly men who run the festival with one female, Miss Mutandi, who is the office administrator. Artists feel that equality must first target the committees that put together and curate the content for the festival. As discussed in previous sections, in terms of the artistic content of the festival, women felt epistemically discriminated against in that there is little work by women or little work that focusses on women's issues at the festival.

Miss Mutandi said there is a need to set up more platforms for women so that they can be encouraged to be a part of the festival. In her own words, these platforms will be spaces "where women can met, engage, network, collaborate, exchange and share ideas" (Runyararo Cynthia Mutandi, in an interview with the writer). However, she insisted that this has to be a collective effort in the arts industry since Intwasa is just one week per year. However, whatever efforts are made, they should be undertaken in such a way that it does not worsen the gendered violence against women at the festival and in the arts industry in general. As argued before, creating a space for women may look progressive in that it can be regarded as creating safe spaces for women, but it might have detrimental effects in creating the impression that women cannot compete with their male counterparts in the mainstream. Whatever women have achieved so far in occupying part of the mainstream art spaces must not be given up, but affirmative action should be directed at assisting them in occupying 50% of the festival. Ms Zulu said she believed that women had to change their attitude, organise and grab the chance to contribute to the growth of the arts and festivals in the city rather than wait for men to give them their salvation in the arts.

Conclusion

According to discussions with artists and administrators of the festival, gendered violence at the Intwasa Arts Festival manifests itself in various ways, including exposure to sexual harassment and abuse. In that the most important and attractive shows that take place at the centre stage of the festival are mostly programmed for late at night in the city centre, it exposes both female artists and audience members to physical bodily harm when they return home to townships and suburbs. This underrepresentation of women in terms of participation in arts, in general, and the festival, in particular, is linked to the fact that women are oppressed in this society. The low numbers of women artists participating at the festival is tied to the gendered epistemic violence of what it means to be a woman and to be an artist in Zimbabwe, in general, and in Bulawayo, in particular. In a context where women face a tough task of self-definition as artists, in navigating the industry, female artists are trying to negotiate a heavily patriarchal social system mostly run and maintained by men in terms of producers, directors, funders and benefactors. The insistence is on merit, which is to say (female) gender and the disadvantage tied to it do not matter, and, therefore, women have to compete with men and prove that they are as talented as them. Such a deceptively fair approach underplays the disadvantage against women, including the challenges they have convincing their families that they can be simultaneously 'artists' and 'women'. Analysis from a postcolonial and decolonial feminisms standpoint – such as anticolonial feminism – makes visible the patriarchy and coloniality of this scenario (Mendoza, 2016, 100). What makes the situation worse is that in all these struggles, at home or in the arts industry, women have to convince men. In a sense, women are heavily subalternised, meaning they cannot speak, and if they speak at all, they cannot be heard (Spivak, 1988). This gendered epistemic injustice at the heart of the participation of women in arts, in general, and the Intwasa Arts Festival, in particular, is such that women are not only defined by men but also spoken for by men, meaning they are spoken *for* and spoken *of*. They are only legible as part of men's speech and not their own voice.

References

Alexander-Floyd, N. (2012). Disappearing acts: Reclaiming intersectionality in the social sciences in a post-black feminist era. *Feminist Formations*, *24*(1), 1–25.

Baya, R. (2016). Awakening city of kings artists. In Norwegian Embassy and Hivos People Unlimited. (2016). *Spotlight on the culture frame in Zimbabwe: Stories from the ground* (pp. 40–41). Retrieved June 15, 2019, from https://knowledge.hivos.org/sites/default/files/spotlight_on_culture_frame.pdf

Bechtold, B. H., Gunn, C., & Hozic, A. (2004). Introduction: Art and political economy. *Review of Radical Political Economics, 36*(4), 461–470.

Bretthauer, MS, B., Zimmerman, T. S., & Banning, J. H. (2007). A feminist analysis of popular music: Power over objectivation of, and violence against women. *Journal of Feminist Family Therapy, 18*(4), 29–51.

Chari, T. (2008). Representation of women in male-produced "urban grooves" music in Zimbabwe. *Muziki, 5*(1), 92–110.

Chivandikwa, N. 2009. Genderized and subversive spectatorship: The case of two Zimbabwean plays. In M. Kolk. (Ed.), *Performing gender in Arabic/African theatre* (pp. 181–194). Amsterdam: Antje von Graevenitz.

Chivandikwa, N., Mhako-Mutonhodza, D., & Sambo, K. (2010). Theatre education and gender equity: Possibilities for the millennium. *Zimbabwe Journal of Educational Research, 22*(2), 181–201.

Collins, P. H., & Bilge, S. (2016). *Intersectionality*. Malden, MA: Polity Press.

Connell, R. W. (1987). *Gender and power*. Cambridge and Oxford: Polity Press and Basil Blackwell.

Connell, R. W., & Messerschmidt, J. W. (2005). Hegemonic masculinity: Rethinking the concept. *Gender and Society, 19*(6), 829–859.

Cooper, B. (2016). Intersectionality. In L. Disch & M. Hawkesworth (Eds.), *The Oxford handbook of feminist theory* (pp. 385–406). New York: Oxford University Press.

Cottle, S. (Ed.). (2000). *Ethnic minorities and the media: Changing cultural boundaries*. Berkshire: Open University Press/McGraw Hill Education.

Fairclough, N. (1995). *Media discourse*. London: Edward Arnold.

Fricker, M. (2007). *Epistemic injustice: Power and the ethics of knowing*. Oxford: Oxford University Press.

Garry, A. (2011). Intersectionality, metaphors, and the multiplicity of gender. *Hypatia, 26*, 826–850.

Giladi, P., & MicMillan, M. (2018). Introduction: Epistemic injustice and recognition theory. *Feminist Philosophy Quarterly, 4*(4), 1–4.

Hall, S. (1997). The work of representation. In S. Hall (Ed.), *Representation: Cultural representations and signifying practices* (pp. 13–74). London: Sage Publications and Open University Press.

Harding, S. (1986). *The science question in feminism*. Ithaca, NY: Cornell University Press.

Hartsock, N. (1983). The feminist standpoint: Developing the ground for a specifically feminist historical materialism. In S. Harding & M. Hintikka (Eds.), *Discovering reality: Feminist perspectives on epistemology, metaphysics, methodology, and the philosophy of science* (pp. 283–310). Dordrecht: D. Reidel.

Inteman, K. (2016). Feminist standpoint. In L. Disch & M. Hawkesworth (Eds.), *The Oxford handbook of feminist theory* (pp. 261–282). New York: Oxford University Press.

Kamara, Y. (2014). Challenges for African women entrepreneurs in the performing arts and designer fashion sectors. In United Nations Educational, Scientific and Cultural Organisation (Unesco) (Ed.), *Gender equality, heritage and creativity* (pp. 108–121). Paris: Unesco.

Karppinen, K. (2008). Media and the paradoxes of pluralism. In D. Hesmondhalgh & J. Toynbee (Eds.), *The media and social theory* (pp. 27–42). London: Routledge.

Krijnen, T., & van Bauwel, S. (2015). *Gender and media: Representing, producing, consuming*. London: Routledge.

Kvale, S. (1996). *Interviews: An introduction to qualitative research interviewing*. London: Sage Publications.

Lugones, M. (2008). The coloniality of gender. *Worlds & Knowledges Otherwise, 2*, 1–17.

McClintock, A. (1995). *Imperial leather: Race, gender and sexuality in the colonial contest*. New York and London: Routledge.

Mendoza, B. (2016). Coloniality of gender and power: From postcoloniality to decoloniality. In L. Disch & M. Hawkesworth (Eds.), *The Oxford handbook of feminist theory* (pp. 100–121). New York: Oxford University Press.

Mlotshwa, K. (2018). "Invisibility and hypervisibility" of "Ndebele women" in Zimbabwe's media. *Agenda, 32*(3), 87–99.

Mohanty, C. T., Russo, A., & Torres, L. (Eds.). (1991). *Third world women and the politics of feminism*. Bloomington and Indianapolis: Indiana University Press.

Oyewumi, O. (1997). *The invention of women: Making an African sense of western gender discourses*. Minneapolis: University of Minnesota Press.

Pohlhaus, Jr., G. (2017). Varieties of epistemic injustice. In I. J. Kidd, J. Medina, & G. Polhaus Jr. (Eds.), *The Routledge handbook of epistemic injustice* (pp. 149–157). London and New York: Routledge.

Sibanda, N. (2018). Theatre training and performance as gender mainstreaming strategies in Zimbabwe: The case of Amakhosi theatre productions. *Theatre, Dance and Performance Training, 9*(1), 68–80.

Smith, V. (1998). *Not just race, not just gender: Black feminist readings*. New York: Routledge.

Spivak, G. C. (1988). Can the subaltern speak? In C. Nelson & L. Grossberg (Eds.), *Marxism and the interpretation of culture* (pp. 271–313). Basingstoke: Macmillan Education.

Spivak, G. C. (2005). Scattered speculations on the subaltern and the popular. *Postcolonial Studies, 8*(4), 475–486.

Tomlinson, B. (2013). To tell the truth and not get trapped: Desire, distance, and intersectionality at the scene of argument. *Signs: Journal of Women in Culture and Society, 38*(4), 993–1017.

Tuana, N. (2017). Feminist epistemology: The subject of knowledge. In I. J. Kidd, J. Medina, & G. Polhaus Jr. (Eds.), *The Routledge handbook of epistemic injustice* (pp. 125–138). London and New York: Routledge.

Webb, J. (2009). *Understanding representation*. Los Angeles, London and New Delhi: Sage Publications.

6 Gender, transgression and sexual violence at Australian music festivals

Bianca Fileborn, Phillip Wadds and Stephen Tomsen

Introduction

Imagine you are in a crowd of tightly packed, sweaty bodies, not-so-patiently waiting for your favourite band to take the stage. You have been anticipating this moment for months – with tickets purchased, outfits meticulously planned, drugs of choice acquired. Your crew of friends have scored a choice camping spot, and so far, the festival has been nothing but fun: drinking, dancing, socialising, hearing some amazing tunes. As tension and excitement grows within the crowd, the mass of bodies surge forwards and back, pressed up against each other so tightly it is impossible to avoid touching and being touched. Somewhere amongst these entangled bodies, a hand reaches through and squeezes you firmly on the backside – too intentional to be 'just' an unfortunate outcome of the crowded conditions and embodied excitement bubbling around you. But maybe you're reading too much into it? You turn around to try and identify the person responsible, but it is impossible to figure out who it might have been – the faces behind you give nothing away, and most are focussed firmly on the stage. Upset and violated, you concede there is little that can be done and try to enjoy the show.

Scenarios such as the previous are seemingly all too common at music festivals across the globe. A wealth of media reporting and activism has begun to generate public discussion on this issue (Davies, 2017; Dmytryschchak, 2016; Lewis, 2017). Yet academic research remains scant. Despite the lack of a robust evidence base, the small body of research that has occurred points to these encounters being routine, normalised and harmful. Recent research out of the UK found that between 30%–40% of young women have experienced sexual harassment (Bows, 2019; YouGov, 2018), with 8% reporting that they experienced sexual assault at a music festival in the past 12 months (Bows, 2019). A more informal survey of women attending Coachella in 2019 found that 1 in 6 experienced sexual harassment and that these incidents were rarely reported (Hayden, 2019). These initial insights match the findings of our own study, which also found that sexual harassment and assault were perceived to be common at Australian music festivals, that it was under-reported when it occurred (and responded to poorly when it was reported) and that it resulted in women withdrawing from or limiting how they participated in the music scene (Fileborn et al., 2019; Fileborn et al., forthcoming).

What is it about music festivals that facilitate the occurrence of sexual violence? Is there anything unique about the festival context that we should take into account in order to fully understand, intervene in and prevent sexual violence in this setting? Or is this just a case of gendered violence as usual, given that (particularly) women and lesbian, gay, bisexual transgender, queer, questioning and others (LGBTQ+) encounter various forms of sexual and other violence across their interpersonal relationships, and in other social and cultural locations, including the street (Vera-Gray, 2016), special events (Tomsen & Markwell, 2009), licensed venues (Fileborn, 2016) and educational institutions (AHRC, 2017)?

The constructed vignette opening this chapter draws together a range of features that commonly appeared in participants' experiences: crowded spaces with frequent incidental touching, a distracted crowd minimising the likelihood of bystander intervention, drug and alcohol consumption, the temporary and transient nature of festival spaces and perpetrators' ability to disappear into the masses. As we discuss in a moment, such features combine with particular cultural, social and gendered norms at festivals and within the music industry at large.

Assemblages of sexual violence

In this chapter, we draw on the overarching concept of assemblage in making sense of how sexual violence occurs at music festivals (Deleuze & Guattari, 1987). We argue that this approach provides a sophisticated and nuanced framework for understanding sexual violence that avoids reducing it to singular categories of causation (i.e. that sexual violence is *only* a matter of structure, or discourse or gender and so forth). According to this approach, we need to view sexual violence as situated within the complex interplay between material/non-human and discursive features of festivals – while these factors also reinforce and sediment the broader norms and structural elements that underpin this violence.

Assemblage thinking is used here to analyse sexual violence occurring at the incident or 'micro-level' – how it unfolds in particular contexts – rather than as a macro-level, broadly applicable explanatory framework. The concept of 'assemblage' has its basis in the work of French philosophers Deleuze and Guattari (1987), though we also refer at times to the closely related actor-network theory developed within science and technology studies (Hayward, 2012; Latour, 2005) and critical-materialist feminism (e.g., Barad, 2003). Drawing on Deleuze and Guattari (1987), Shaw described assemblages as "about arranging co-existences: about how the stable and the unstable, the solid and the light interact to produce the thing" (2014, 88). Assemblages are emergent and fluid interconnections (or *intra*-actions – Barad, 2003) between material and discursive, human and non-human elements. In this regard, assemblage theory challenges the notion that "objects, spaces and places are ... 'passive backdrops'" to human interactions, instead suggesting that "socio-spatial settings ... should be included in investigations" (Pedersen et al., 2017, 165).

While assemblage thinking decentralises the importance of human action – taking a 'rhizomatic' approach that flattens out relationships between human/non-human elements – this does not mean that humans are passive or inert (Dilkes-Frayne, 2016; Hayward, 2012; Pedersen et al., 2017). However, assemblage thinking takes a radically different ontological approach, one that recognises "the agency of nonhuman, or 'more-than-human' forces" (Duff, 2016, 16). Rather than focussing exclusively on agentic human action, we must instead ask "how action ... is generated in encounters" by looking at "*all the factors* ... that mediate or transform a given phenomenon" (Duff, 2016, 16, original emphasis). In contrast to the notion that humans are bounded and autonomous, we are instead constituted through complex entanglements or 'folds' with the material and discursive worlds. As Malins et al. (2006) explained, the Deluezian concept of folding accounts for the mutually constitutive intra-relationships between space, discourse, self-governance and subjectivity. Importantly, this concept of 'folding' retains space for "transformative change, and for the creation of the new" (Malins et al., 2006, 512). Assemblage thinking thus provides a framework that allows us to take seriously the intra-action between material, human and discursive worlds while avoiding the essentialist trap of suggesting that material and non-human features directly *cause* or determine our actions (Malins, 2004).

Our use of assemblage stems in part from perceived limitations to discursive, cultural and structural approaches that dominate theoretical understandings of sexual violence, productive as these are. Sexual violence has, first and foremost, been theorised as an expression of, and a means of (re)producing, systems of gender-based power – and, certainly, power relations are central (if perhaps insufficient in and of themselves) in understanding the emergence of particular assemblages (Duff, 2016). However, post-structuralist accounts recognise the centrality of discourse in constituting sexual violence: there is no underlying stable or essential understanding of what sexual violence 'is' (Heberle & Grace, 2009). Rather, our definitions and understandings of sexual violence shift and change over time and place, with the boundaries of what 'counts' being fluid and contested (see, for example, Fileborn & Phillips, 2019). Dominant discursive constructions of sexual violence present important implications for how survivors understand and label their experiences (and whether this label is *recognised* by others), with discourse allowing us to articulate certain experiences as harmful (Gunnarsson, 2018). Yet, in other respects, a purely discursive approach is unsatisfying given that sexual violence also involves very 'real' things happening to and through our bodies (see Fileborn, 2016). Sexual violence has a material, embodied 'reality' to it, as much as our understandings of these corporeal happenings are shaped and delimited through discourse.

Additionally, this violence occurs within particular spaces and places – something that has been largely ignored in existing scholarship beyond challenging the misconception that 'real' sexual violence is perpetrated in public space at the expense of that occurring in the 'private' or domestic realm (Fileborn, 2016). That space/place and geographic features may be *productive* of violence has only occasionally been considered in relation to sexual violence (Fileborn, 2016).

This reflects a broader trend in criminological thinking, which has tended to approach the issue of space and geography in a reductive and, at times, overly deterministic way (Hayward, 2012). As Hayward explained, criminologists have traditionally approached "the environment simply as a geographic site and not as a product of power relations, cultural and social dynamics'" (2012, 441). Yet, as scholars such as McDowell argued, spaces "have a constitutive effect on social processes" (1996, 28) – though this is not to say that spaces are produced or experienced in any one stable way (see, e.g., McDowell, 1996; Valentine, 1996).

In using assemblage thinking, we ask how these cultural, discursive, spatial, material and human elements come together in fluid and temporally specific ways to generate incidents of sexual violence. This approach does not intend to downplay or ameliorate the active choices made by, and responsibility of, perpetrators of sexual violence. Rather, it aims to examine how these choices and actions are situated within a broader interplay (or *intra*-play) of disparate factors. We suggest that this approach is helpful in thinking through how sexual violence occurs *in the moment*, working alongside higher-level macro-theories relating to power, gender and discourse (Duff, 2014).

'Assemblage thinking' has been applied to other aspects of sexual violence and rape culture (Fanghanel, 2019), music festivals and the night-time economy, particularly drug and alcohol consumption and 'risk', and sexual interaction (Dilkes-Frayne, 2016; Malins, 2004; Pedersen et al., 2017). Ella Dilkes-Frayne's (2016) work provides a key example of how assemblage thinking (or actor-network theory more specifically) can be productive in thinking through the relationships between festival space and drug consumption. For instance, Dilkes-Frayne's (2016, 32) analysis demonstrates how the spatial arrangements of festival campsites lend themselves to the buying, selling and consumption of drugs on account of their being simultaneously 'open and porous' yet private spaces.

Music festivals, sociocultural norms and sexual violence

Music festivals are dynamic and diverse social and cultural sites. While it is not possible to make any definitive statement on the norms and practices associated with all festivals (see Dilkes-Frayne, 2016), there are, nonetheless, some common themes apparent in the literature to date that hold particular relevance to our study. We suggest that festivals can – to some extent – be conceptualised as liminal, 'carnivalesque' and transgressive spaces (see also Anderton, 2011; Jaimangal-Jones et al., 2010; Pielichaty, 2015; Riches, 2011). They are regularly framed and experienced as a 'time out', holiday or retreat from the routines of everyday life, where restrictive regulatory forms and structures increasingly limit traditional outlets for social and cultural expression (see Presdee, 200; Morrissey, 2008; Hackley et al., 2013). Here collective forms of transgression lubricate social relations, enable deeper immersion in the festival experience and temporarily unshackle (at least ostensibly) festival patrons from external pressures while simultaneously facilitating the performance of desired social identities (see, e.g., Dilkes-Frayne, 2016; Pielichaty, 2015).

The transgressive and carnivalesque features of festivals are perhaps best illustrated through excessive and determined consumption of AODs. While there are substantive differences across festival genres in consumption type and extent, it is fair to say that these are sites of heightened substance use. Australian research illustrates that young patrons attending a single-day music festival consumed AODs at rates that exceed those of their non-festival attending peers (Lim et al., 2008, 2010). In our own study, drug and alcohol consumption were common, with virtually all participants consuming alcohol (99%), most at what would be considered 'risky' levels (Fileborn et al., 2019). Just under half of participants said that they took other drugs (47.8%) – including illicit substances and non-prescribed use of pharmacy medications – with a large minority engaging in poly-drug use (31.5%).

To understand sexual violence at festivals, it is important to consider the *gendered* production of festival space, culture and social norms, as well as situating this within the broader Australian music industry. Pielichaty's (2015, 246) ethnographic study of gender at a UK-based music festival is instructive here, with their work illustrating that "hegemonic masculinity dominates the festival scene" and that gendered power relations are reproduced within festival sites despite the perception of festivals as 'transgressive' spaces. Again, while we resist making any definitive claims about the 'nature' of the Australian music scene – or even suggest that there is one coherent 'scene' to speak of – research to date, nonetheless, consistently illustrates that the industry is male dominated on multiple fronts. Cooper et al.'s (2017) analysis of gender inequality in the industry, for example, shows that men enjoy greater access to power, resources and cultural kudos at virtually all levels. Men occupy significantly more managerial and leadership positions, dominate festival line-ups, radio play and awards, and they receive more pay for the work they do. Male musicians are constructed as possessing greater technical skill and knowledge, while women tend to be typecast as the 'sexy' frontwoman (Bannister, 2006; Davies, 2001; Fileborn et al., 2019; Strong, 2011, 2014). In all likelihood, LGBTQ+ people, people of colour and others occupying marginalised positions also face exclusion, discrimination and various inequalities – as Waitt has suggested, festivals can be spaces in which "social hierarchies and normative ideas are reinforced rather than inverted" (2008, 515). This heterosexual, masculine dominance forms part of the broader cultural backdrop that facilitates and enables sexual violence at festivals (Fileborn et al., 2019): it is a key component of the assemblage, though these conditions are also emergent, unstable and fluid.

A final aspect to consider here are the physical, geographic and temporal features of festival sites. Festivals are temporary and liminal spaces: for some events, entire cities and infrastructure may be put together and then pulled apart within weeks (Jaimangal-Jones et al., 2010). Some long-standing events may have access to more stable infrastructure – for example, the festival we conducted observations at had 'permanent' shower and bathroom facilities built on-site. In Australia, camping festivals are often located in rural areas or national parks. The festival may span a single day, several nights or even weeks. Camping sites are often haphazardly assembled (Dilkes-Frayne, 2016), with varying levels of direction from festival staff across events. Temporal, cultural and environmental features shape patterns of drug and alcohol consumption, as well as the opportunity (and often explicit intention) to engage in consensual sexual activity (Dilkes-Frayne, 2016; Fileborn et al., 2019). All of this means that music festivals present unique, temporary and fluid spaces that create potential implications for producing sexual violence.

Methods

To examine how these diverse and unique features of music festivals come together in fluid and temporally specific ways to facilitate sexual violence, we draw on narratives from 16 qualitative, semi-structured interviews conducted with women who have experienced sexual violence at Australian music festivals (n = 13) and individuals who have either witnessed or responded to sexual violence at festivals (n = 3). These interviews formed one component of a larger pilot study – the first of its kind within Australia. The larger study included an online survey with 500 patrons who attended a large camping festival in 2017–2018 and on-site observations undertaken by two of the authors.

Interview participants were recruited from across Australia, and we were interested in speaking to individuals who had experienced sexual violence (self-defined) at *any* Australian music festival. Recruitment was also extended to include individuals who had responded to or witnessed sexual violence at festivals. Although we sought to recruit participants from diverse backgrounds, the majority were (unsurprisingly) women, Anglo-Australian and university educated and/or employed in professional jobs. Participants were recruited through a range of avenues, including project social media accounts, and media articles. The interviews were designed to undertake an in-depth exploration of participants' experiences and addressed topics that included reasons for attending festivals; participants general sense of safety at festivals; experience(s) of sexual violence, including location, perpetrator identity, type(s) of violence, disclosure and reporting and impacts of the experience; and how participants would like festivals to respond to sexual violence. Typically, interviews took between 50–70 minutes and were completed in person, via Skype or by phone, depending on participants' preferences and locations. All interviews were digitally recorded and transcribed by a professional service. Participants were able to request a copy of their transcript to check and provide feedback to the research team post-interview. Undertaking research on sexual violence is, of course, ethically sensitive, and a range of strategies were put in place to support participants and afford them control over the interview process. In addition to the strategies already mentioned, participants were provided with a copy of the interview schedule before deciding whether to take part and could let the researcher know at the beginning of the interview if there were any questions they did not want to discuss. Likewise, participants were provided with an information statement detailing the nature and focus of the study and potential risks of participating. Finally, all interviews closed with a debriefing chat, and participants were provided with contact details for sexual assault support services.

Interview data were analysed using thematic analysis (TA). The approach we took to TA used both inductive and deductive coding, with codes developed based on recurrent themes we identified and by using the interview question areas as pre-determined codes. In conducting the analysis, we were mindful of commonalities and patterns across participants' experiences but also paid attention to key points of difference, divergence and contradiction. In line with the theoretical framework guiding this research, we do not view the results presented here as 'generalisable' or transferable to other contexts. Rather, we view these findings as partial and situated, produced within a specific context and for a particular purpose.

Assemblages of sexual violence at festivals

In the following sections, we examine the experiences of three participants in-depth in order to draw out the assemblage of factors that come to play in enabling sexual violence. We aim to illustrate the complex alignment of material, human and non-human components that create the possibility of sexual violence and harassment occurring and suggest that we cannot reduce these encounters to any one factor within the assemblage.

Camilla

Camilla recounted numerous experiences of sexual violence and harassment in her interview; however, we examine one encounter in detail here. Camilla shared that she often feels "overstimulated in big crowds ... when there are lots of lights and things". She was attending a camping festival and needed to retreat to her tent to get away from the overwhelming environment of the stage area. Camilla's friends elected "to stay up and party", taking advantage of – and accentuating – the hedonistic festival culture. A male friend escorted Camilla back to their campsite, and she soon found herself in the liminal state between awake and sleep. She heard someone enter her tent, but assumed it was one of her friends. However, after this individual spent several minutes noisily "rustling around" the tent, Camilla realised it was a stranger. Camilla asked him what he was doing in the tent, and the stranger asked who she was. He was evidently intoxicated and had wandered into the wrong tent, assuming it was his.

Rather than leaving, the stranger decided to initiate conversation with Camilla. Despite feeling ill at ease, she felt unable to ask him to leave, "because you never know how that is going to go when you're a girl" and engaged in prolonged conversation with him. He attempted to flirt, and at one point asked, "Can I share your sleeping bag ... because I don't know where to go?" Fortunately, one of Camilla's friends eventually returned to the campsite and asked the stranger to leave. While she now views the experience through a lens of "dark humour", she also reflected that it was a "potentially dangerous situation", particularly as it was relatively early in the night before the headlining act had played, meaning that the campsite was deserted. As Camilla reflected,

> It was almost this eerily quiet environment and I was like I don't know if there is security wandering around here? Are they all in at the festival [as opposed to the peripheral camping areas]? The funny thing about it is that when you can't remember the last time that you saw a security guard when something like that happens and you think like okay if I yell out, is it going to be a group of his friends that come? ... what's actually going to be the end result of what happens?

While Camilla had no way of knowing if the stranger was genuinely lost or had intentionally entered her tent, she felt it was likely he was lost but had decided to stay because of "this mentality that guys have ... going to festivals ... they like to have the stories of it. And they will get drunk and be like 'remember that one time that you ...'".

There was a range of issues at play that come together to produce Camilla's experience and enable the actions of the young man in this encounter. It may be tempting to dismiss the example as one in which "nothing really happened" (Kelly & Radford, 1996) – a discursive position that Camilla herself draws on in suggesting it was "a *potentially* dangerous situation". However, this incident had a profound impact on Camilla, to the point where she opted to pay substantially more money at the next camping festival she attended to sleep in a more 'exclusive' section of the festival, believing it would be more controlled (though this, unfortunately, did not prevent Camilla from experiencing further harassment). This points to the importance of taking this experience seriously as sexual harassment and as situated on the continuum of sexual violence.

Aspects of the physical environment interplayed with Camilla's embodied experiences of the space. Camilla found the festival environment overwhelming, with the confluence of bodies, lights and noise encountered as sensory overload. The coming together of these environmental features and corporeal responses directed Camilla through the festival space to a comparatively more isolated and quiet location where she could seek temporary refuge. As our own observations of a camping festival confirmed, camping sites are vast and often deserted spaces, especially during peak performance times. The sea of tents – often of similar colours and sizes, combined with an absence of signage or way-finding devices – can quickly become disorienting and confusing, something that was no doubt heightened by whichever substances the stranger had imbibed. This is further compounded by the lack of lighting in the night-time, with these environmental features combining to make locating one's tent a challenging task. The risk of a stranger either intentionally or unintentionally accessing a tent is made possible by the ease of access, especially when locks are so rarely used in these settings. Such material and environmental features created the potential for strangers to 'accidentally' enter the wrong tent. We heard several other examples where this had happened to our interview participants or their friends, though in less threatening contexts.

The spatial arrangement of the campsite reflects Dilkes-Frayne's (2016) observation of their "open and porous" yet private nature. However, in this case, rather than generating contexts for the consumption of illicit drugs, this environment may generate the potential for sexual harassment/violence to unfold.

Camilla's sense of unease and isolation was further compounded by the apparent lack of security staff and formal surveillance. In comparison to other festival spaces where there was an overwhelming mass of bodies, the campsite was sparsely populated and seemingly not subject to routine observation and patrols (something that was again noted in our own observations of a festival campsite). The practices of festival security, and their absence from this particular space, further contributed to the creation of an isolated space – an environmental setting that may have facilitated or promoted potential perpetration. This absence may have been temporal, with no perceived need to have security circulating the campsite at a time when the vast majority of patrons were concentrated around the main festival stages (though we can only guess as to whether this was actually the case in the absence of information as to how security decided where to direct staff members).

The transgressive and carnivalesque festival culture also impacted on Camilla's experience. Notably, Camilla's isolation was in part a result of her friends continuing to 'party' in the main festival area. This is not to assign blame to her friends but rather to illustrate how the vast environment and spatial segregation of activities contributed to Camilla being alone. More substantively, the stranger's intoxication seems to have played a contributing factor to his disorientation and struggle to successfully navigate his way back to the correct campsite. It is the entanglement between the environmental context and his inebriated state that directed his body towards the wrong tent.

Finally, a range of discourses relating to festivals, gender and sexual violence intersected in producing what happened once the stranger had found his way into Camilla's tent. Several dominant discursive positions pertaining to gender 'fold' into the embodied actions of Camilla and the stranger (see Malins et al., 2006 for an overview of the Deleuzian concept of 'folding'). Camilla's reluctance in telling the stranger to "fuck off" draws on the trope of "stranger danger" – that women are most "at risk" of sexual violence from unknown men in dark and isolated public spaces (conditions that largely reflect those that Camilla was in). Again, this is in no way intended to place blame or responsibility on Camilla but rather to illustrate how this particular discourse played into the construction of this situation as a dangerous one and the ways in which it may have delimited her potential for action. Camilla's perceived need to delicately manage the situation further demonstrates the impossible double bind that women often experience when negotiating the threat (or actual occurrence) of men's violence. Women are held responsible if their direct or blunt response 'provokes' the perpetrator while also being deemed responsible for failing to clearly express their lack of interest. This is further complicated by the fact that women may also be accused of 'over-reacting' and over-estimating the level of threat posed by a particular man (see Vera-Gray, 2018).

In the case of the stranger, it seems that his actions may have been folded into particular norms of masculinity specific to the festival environment (at least based on Camilla's interpretations of his actions). His decision to 'try his luck' with Camilla after accidentally stumbling into her tent may have been underpinned by the hedonistic and transgressive cultural environment: by the construction of festivals as a space where (to a limited extent) 'anything goes' and the potential for spontaneous sexual hook-ups presents as a possibility. There are also homosocial imperatives apparent here. The stranger's actions reflect Grazian's (2007) concept of the "girl hunt", whereby groups of young, heterosexual men attempt to 'pick up' or flirt with women in order to (re)affirm their status as suitably 'masculine' with their male peers. In other words, sexual conquests or their attempt function as a type of hegemonic masculine performance undertaken for an audience of male peers. While the stranger's peers were not physically present to observe this attempt at 'hooking up' with Camilla, it may have served this homosocial function later by providing material for nostalgic storytelling. Indeed, such storytelling was documented on numerous occasions during the fieldwork component of this project. Here groups of young men could routinely be heard loudly recounting stories of their (real or attempted) sexual conquests. Often in proximity to major thoroughfares, these performances were evidently not only intended for their immediate audience but for all others in the immediate vicinity, presumably in the hope that their masculine status would be suitably enhanced by their stories of domination and sexual achievement.

Penny

Similar to Camilla, Penny had numerous experiences of harassment – in fact, she said, "It is definitely something that has happened to me at almost every festival". While Penny felt she could usually "brush off" these incidents, she discussed one recent experience that had a more overt impact on her. Penny had volunteered at the festival and so had attended alone rather than with a group of friends. She made friends with another female volunteer during their shift, and they made plans to watch a band together later, meeting up with a group of people who the other volunteer knew but was not particularly close friends with. Penny recalled that the crowd was "quite rowdy", which she attributed in part to the "rapper" on stage. They were positioned towards the front of the stage in the thick of the mosh pit, where the group was somewhat separated in the crowd but still in the same general area.

The crowd was pushing forward, and Penny turned around to see what was happening. The man standing immediately behind her reassured her that he was not intentionally pushing her and that he was also being pushed by the crowd. However, moments later, Penny said she could feel "someone's hands touching my lower area around my butt and my hips and stuff just caressing". Penny questioned whether she was "just imagining it", but as she was completely sober, she felt reassured that this was not the case and tried to mentally block out what was happening. Despite her perpetrators' actions, she did not feel unsafe because the man had been friendly and "seemed like a genuine person". This made it more difficult for Penny to process the nature of his actions at the time, with their impact not hitting her until she was home later in the evening. Because Penny did not know the group she was with well, she felt unable to communicate what was happening to them and said, "They didn't have an obligation to intervene".

It is interesting that Penny felt unable to simply 'brush off' this experience as she had with others in the past. There may be a number of emergent factors that shaped this experience. Discursive constructions of 'real' or 'serious' sexual violence may have contributed. Dominant discourses tend to position iterations of violence involving physical touch as 'more serious', although it was unclear from Penny's discussion whether this incident was more physically invasive than others she had experienced. Penny's inability to 'brush off' this incident may also reflect the cumulative nature of these 'small' encounters (see Fileborn & Vera-Gray, 2017). As this was the most recent in a long line of harassing experiences, it may have been the collective weight of these that contributed towards the increased harm of this particular incident. This also illustrates that assemblages are not necessarily entirely unstructured or free-forming entities, disconnected from broader structures of power, lived memory and so forth. Though assemblages are emergent and fluid, replete with potential and becoming, they are not unbounded. In other words, these assemblages are forming within 'striated' spaces, which, as Alexandra Fanghanel explained are "homogenised, ordered, territorialised space[s]" (2019, 5).

The spatial context was a prominent feature of this assemblage, and the crowded mosh pit area was a common feature in participants' experiences. From Penny's perspective, the 'rowdiness' of the crowd was generated through the genre of music being played on stage (see also Riches, 2011). There was an apparent 'folding' of the cultural norms of the musical style with the embodied actions of the crowd (though this is by no means limited or specific to rap music). Collective energy ripples through the mass of bodies, pushing the crowd against each other, leading to inevitable touching. This is not inherently problematic, and the energy, embodied expression and unpredictability associated with the 'mosh pit' can be a core attraction of music festivals, as well as providing a liminal space in which women might contest dominant norms of femininity (Fileborn et al., 2019; Riches, 2011). Normative touching in the mosh pit caused Penny to at first question her interpretation of what was happening – that she might just have been 'imagining' or over-interpreting matters. However, the *absence* of alcohol or other drug consumption enabled Penny to confidently interpret and affirm her lived experience. Simultaneously, the crowded space provided camouflage for the perpetrator, with others in the crowd unable to see what was happening amongst the mass of jammed bodies.

For Penny, the social context featured as a key aspect of her experience. While Penny typically went to festivals with groups of friends, in this case, her role as a volunteer meant that she was on her own. She was socially isolated. As a result, she felt unable to speak out during the incident. The social context also shaped

Penny's expectations of bystander intervention, with her social distance from the group she was with allaying any perceived responsibility for them to provide support or intervene in any way. While the absence of intervention did not *cause* the perpetrator's actions, it may have enabled them to continue and prevented any informal or formal ramifications. More broadly, a lack of both bystander intervention and repercussions for the perpetrator may also enable and entrench this behaviour as 'normal' at festival events. A lack of social familiarity may have prevented other members of the social group from recognising that anything was wrong on account of their lack of knowledge regarding Penny's 'usual' way of acting. Finally, social norms appear to have allowed Penny to feel 'safe' in the presence of the man who had just physically violated her, because he was 'genuine' and 'friendly'. Penny's sense of safety here reflects the discursive construction of perpetrators as 'monsters', with her 'nice' perpetrator located outside of this and, therefore, framed as non-threatening in this context.

Angela

Angela also recalled numerous examples of sexual harassment and violence at festivals. The incident we draw on here was one that Angela witnessed at a large, single-day festival some years ago. One of the headlining bands was on stage, playing to a packed crowd. During the band's set, women were sitting on top of men's shoulders – perhaps to get a better view of the stage and to demonstrate their enthusiasm for the band. Large screens on either side of the stage projected footage of the band but also occasionally cut to show audience members. Angela noticed that the "camera would zoom in on … these girls that were on top of these guys shoulders". After this had happened a number of times, the crowd started "cheering and basically pressuring these girls … to take their top off". While some women were seemingly happy to oblige, one refused to do so "and the crowd booed … and eventually I think she did take her t-shirt off". Angela described herself as being a "very hot-headed feminist at the time" and found this spectacle of mass coercion upsetting and started to cry out of anger "about the inequality". Angela's friends dismissed her concerns with the claim that "it's fine, it's music, this is what happens, they want to do it".

In this incident, we can see a complex assemblage of technology, music industry norms and the mass of bodies in the crowd aligning to create the potential for sexual harassment to take place. The technological affordances of the camera allow images of the crowd (and band) to be transmitted on to large screens, visible to the audience. Of course, the humans operating this technology have also used the cameras to repeatedly focus on women in the crowd. In fact, Angela recalled that the crowd's response seemed to directly influence the focus of the camera operator, with these elements folding together with the camera:

> I am sure that the camera would have zoomed in on one girl like just a crowd shot the first time but they continued to do it when they realised it was getting that reaction.

This has merged with the gendered practices discussed earlier in the chapter that construct women in the music scene as hyper-sexual and as, in all likelihood, only being present for the sexual gratification or pleasure of men. Indeed, this is reflected in the dismissive response of Angela's friends – that such an incident was to be expected because "it's music". This was further reinforced for Angela during an experience at the same festival some years later when she was invited to an after-party and expressly told to bring only female friends. Angela viewed this as situated within "the whole ingrained sexist, womaniser, rock and roll sort of vibe" of the broader music industry, which demonstrates how assemblages may be shaped (but not wholly determined) by broader structural elements (Duff, 2016). It is likely that Angela has (re)interpreted her understanding of this incident over time and in light of subsequent experiences illustrating how this particular assemblage continues to emerge and unfold over time.

Further demonstrating the presence and effect of these masculine norms was the group response to the projected images. The crowd booing and pressuring the non-conforming woman to take her top off and expose her breasts not only shows the interconnectedness of various environmental features of the front stage area but also reinforces to the audience that the festival environment is a masculine one where women's bodies are objects for entertainment and sexual gratification. While this may sometimes be seen as a fun part of the liminal experience of festival events, it is clear from this example that these unruly and carnivalesque moments are often morally dubious and ought not escape critique (see also Pielichaty, 2015).

Consumption of alcohol again featured in Angela's experience. She recalled that at the time, she would drink very heavily at festivals. Because of the inflated prices of water and other non-alcoholic beverages, "a bottle of water costs as much as any kind of alcoholic drink, so you would kind of just drink". Angela was heavily intoxicated when she witnessed the incident discussed here, so she did not "know exactly if I remember 100% clearly or not". In contrast to Penny's experience, Angela's state of inebriation caused a level of uncertainty and doubt over the nature of "what happened". Her embodied and physiological state contributed towards the production of an inherently unstable assemblage. Angela's comments also draw attention to the role that capitalist imperatives and the economic structure of festivals can play in producing particular consumption practices (see also Anderton, 2011), with these structural elements thus folding into intoxicated corporeal states.

Conclusion

In this chapter, we set out to explore the potential of 'assemblage thinking' in understanding how individual incidents of sexual violence unfold in music festival settings. Our in-depth analysis illustrates how a complex range of material, discursive, human, non-human factors come together to facilitate sexual violence; these individual components of the assemblage cannot be readily disentangled from one another. Notably, the physical environment (such as tents, crowded spaces), norms relating to gender, festivals and sexual violence and alcohol and drug consumption were common features in participants' experiences, with these features 'folding' or intertwining with each other such that they were mutually constituted and unable to be disentangled into their component parts.

While festivals may be conceived of as 'transgressive' and carnivalesque spaces, our analysis suggests that festival assemblages emerge in highly 'striated' – or ordered – ways to return to Deleuze and Guattari (1987) that, by and large, work to sediment and reinforce factors underpinning sexual violence (see also Fanghanel, 2019). This challenges the notion that festivals are particularly 'transgressive' spaces. Such findings reflect Pielichaty's (2015, 236) argument that, while festivals are "positioned within the supposed status-free refuge of liminality" and carnival, they in fact continue to be striated spaces in which gendered norms are often heavily reinforced (though this plays out in highly complex and nuanced ways; see also Riches, 2011). However, while festival spaces are heavily striated ones, assemblages are always immanent and in a process of becoming. As such, there is potential to disrupt the assemblages giving rise to the possibility of sexual violence at festivals. While continuing to attend to the 'macro-level' structural factors underpinning sexual violence, we must also assess assemblages "in relation to ... [their] specific ethical and political tendencies: *Does it open up potential movements and transformations?* Or, *Does it block new connections and becomings?*" (Malins, 2004, 485, original emphasis). Currently, festivals arguably fall in to the latter category. As such, we need to question how we can create "smooth spaces" that are "not predetermined by structures or categories but which are open to creative movement in any direction" (Malins, 2004, 486). However, in preventing sexual violence, we likely require processes of de- and re-territorialisation. In some festival spaces, particularly campsites, striation may in fact be *required* in order to better prevent and respond to this violence: it is a lack of regulation and norms in this case that is at issue. Likewise, prevention may require the development of spaces that are striated in ways that *do not* enable or facilitate sexual violence to occur or that normalise and excuse it when it does. Nonetheless, it is the constant process of becoming, and potential for transformative change, that opens up possibilities for preventing sexual violence at festivals. The challenge now is to harness this potentiality to move towards new ethical futures.

References

Anderton, C. (2011). Music festival sponsorship: Between commerce and carnival. *Arts Marketing: An International Journal, 1*(2), 145–158.

Australian Human Rights Commission. (2017). *Change the course: National report on sexual assault and sexual harassment at Australian universities.* Sydney: AHRC.

Bannister, M. (2006). Loaded: Indie guitar rock, canonism, white masculinities. *Popular Music, 25*(1), 77–95.

Barad, K. (2003). Posthumanist performativity: Toward an understanding of how matter comes to matter. *Signs: Journal of Women in Culture and Society, 28*(3), 801–831.

Bows, H. (2019). How can sexual assaults at festivals be stopped? *BBC*. Retrieved July 6, 2019, from www.bbc.com/news/uk-48447964

Cooper, R., Coles, A., & Hanna-Osborne, S. (2017). *Skipping a beat: Assessing the state of gender equality in the Australian music industry*. Sydney: The University of Sydney Business School.

Davies, H. (2001). All rock and roll is homosocial: The representation of women in the British rock music press. *Popular Music, 20*(3), 301–319.

Davies, H. (2017). Are music festivals doing enough to tackle sexual assault? *The Guardian*, July 25. Retrieved April 12, 2018, from www.theguardian.com/lifeandstyle/2017/jul/25/music-festivals-sexual-assault-rape-safe

Deleuze, G., & Guattari, F. (1987). *A thousand plateaus: Capitalism and schizophrenia*. Minneapolis: University of Minnesota Press.

Dilkes-Frayne, E. (2016). Drugs at the campsite: Socio-spatial relations and drug use at music festivals. *International Journal of Drug Policy, 33*, 27–35.

Dmytryschchak, G. (2016). Man charged with urinating on woman at Spiderbait gig. *The Age*, November 19. Retrieved March 8, 2018, from www.theage.com.au/national/victoria/man-charged-with-urinating-on-woman-at-spiderbait-gig-20161119-gst72h.html

Duff, C. (2014). The place and time of drugs. *International Journal of Drug Policy, 25*, 633–639.

Duff, C. (2016). Assemblages, territories, contexts. *International Journal of Drug Policy, 33*, 15–20.

Fanghanel, A. (2019). *Disrupting rape culture: Public space, sexuality and revolt*. Bristol: Bristol University Press.

Fileborn, B. (2016). *Reclaiming the night-time economy: Unwanted sexual attention in pubs and clubs*. London and New York: Palgrave Macmillan.

Fileborn, B., & Phillips, N. (2019). From "me too" to "too far"? Contesting the boundaries of sexual violence in contemporary activism. In B. Fileborn & R. Loney-Howes (Eds.), *#MeToo and the politics of social change*. London and New York: Palgrave Macmillan.

Fileborn, B., & Vera-Gray, F. (2017). "I want to be able to walk the street without fear": Transforming justice for street harassment. *Feminist Legal Studies, 25*(2), 203–227.

Fileborn, B., Wadds, P., & Barnes, A. (2019). Setting the stage for sexual assault: The dynamics of gender, culture, space and sexual violence at live music venues. In C. Strong & S. Raine (Eds.), *Towards gender equality in the music industry: Education, practice and strategies for change* (pp. 89–102). New York: Bloomsbury.

Fileborn, B., Wadds, P., & Tomsen, S. (2019). *Safety, sexual harassment and assault at Australian music festivals: Final report*. Sydney: University of New South Wales.

Fileborn, B., Wadds, P., & Tomsen, S. (forthcoming). *Sexual harassment and violence at Australian music festivals: Reporting preferences and experiences of festival attendees*. *Australian and New Zealand Journal of Criminology*. Accepted for publication 7 January 2020.

Grazian, D. (2007). The girl hunt: Urban nightlife and the performance of masculinity as collective activity. *Symbolic Interaction, 30*(2), 221–243.

Gunnarsson, L. (2018). "Excuse me, but are you raping me now?" Discourse and experience in (the grey areas of) sexual violence. *NORA: Nordic Journal of Feminist and Gender Research, 26*(1), 4–18.

Hackley, C., Bengry-Howell, A., Griffin, C., Mistral, W., Szmigin, I., & Tiwsakul, R. A. H. N. (2013). Young adults and "binge" drinking: A Bakhtinian analysis. *Journal of Marketing Management, 29*(7–8), 933–949.

Hayden, N. (2019). One in six women told us they were sexually harassed at Stagecoach, Coachella 2019. *Desert Sun*. Retrieved July 6, 2019, from www.desertsun.com/story/life/entertainment/music/coachella/2019/05/17/1-6-women-sexual-harassment-stagecoach-coachella-2019/1188482001/?utm_source=share&utm_medium=ios_app

Hayward, K. J. (2012). Five spaces of cultural criminology. *British Journal of Criminology, 52*, 441–462.

Heberle, R. J., & Grace, V. (2009). Introduction: Theorizing sexual violence: Subjectivity and politics in late modernity. In R. J. Heberle & V. Grace (Eds.), *Theorizing sexual violence* (pp. 1–13). New York and Oxon: Routledge.

Jaimangal-Jones, D., Pritchard, A., & Morgan, N. (2010). Going the distance: Locating journey, liminality and rites of passage in dance music experiences. *Leisure Studies, 29*(3), 253–268.

Kelly, L., & Radford, J. (1996). "Nothing really happened": The invalidation of women's experiences of sexual violence. *Critical Social Policy, 10*, 39–53.

Latour, B. (2005). *Reassembling the social: An introduction to actor-network theory*. New York: Oxford University Press.

Lewis, M. (2017). The growing epidemic of sexual harassment at Aussie music festivals. *SBS*, October 26. Retrieved April 12, 2018, from www.sbs.com.au/news/the-feed/the-growing-epidemic-of-sexual-harassment-at-aussie-music-festivals

Lim, M. S. C., Hellard, M. E., Hocking, J. S., & Aitken, C. K. (2008). A cross-sectional survey of young people attending a music festival: Associations between drug use and musical preference. *Drug and Alcohol Review, 27*(4), 439–441.

Lim, M. S. C., Hellard, M. E., Hocking, J. S., Spelman, T. D., & Aitken, C. K. (2010). Surveillance of drug use among young people attending a music festival in Australia, 2005–2008. *Drug & Alcohol Review, 29*, 150–156.

Malins, P. (2004). Body-space assemblages and folds: Theorizing the relationship between injecting drug user bodies and urban space. *Continuum, 18*(4), 483–495.

Malins, P., Fitzgerald, J. L., & Threadgold, T. (2006). Spatial "folds": The entwining of bodies, risks and city spaces for women injecting drug users in Melbourne's Central Business District. *Gender, Place & Culture, 13*(5), 509–527.

McDowell, L. (1996). Spatializing feminism: Geographic perspectives. In N. Duncan (Ed.), *BodySpace: Destabilizing geographies of gender and sexuality* (pp. 28–44). London and New York: Routledge.

Morrissey, S. A. (2008). Performing risks: Catharsis, carnival and capital in the risk society. *Journal of Youth Studies, 11*(4), 413–427.

Pederson, W., Tutenges, S., & Sandberg, S. (2017). The pleasures of drunken one-night stands: Assemblage theory and narrative environments. *International Journal of Drug Policy, 49*, 160–167.

Pielichaty, H. (2015). Festival space: Gender, liminality and the carnivalesque. *International Journal of Event and Festival Management, 6*(3), 235–250.

Presdee, M. (2000). *Cultural criminology and the carnival of crime*. London: Routledge.

Riches, G. (2011). Embracing the chaos: Mosh pits, extreme metal music and liminality. *Journal for Cultural Research, 15*(3), 315–332.

Shaw, R. (2014). Beyond night-time economy: Affective atmosphere of the urban night. *Geoforum, 51*, 87–95.

Strong, C. (2011). Grunge, Riot Grrrl and the forgetting of women in popular culture. *The Journal of Popular Culture, 44*(2), 398–416.

Strong, C. (2014). All the girls in town: The missing women of Australian rock, cultural memory and coverage of the death of Chrissy Amphlett. *Perfect Beat, 15*(2), 149–166.

Tomsen, S., & Markwell, K. (2009). *When the glitter settles: Safety and hostility at and around gay and lesbian public events*, Research and Public Policy Series, No 100. Canberra: Australian Institute of Criminology.

Valentine, G. (1996). (Re)negotiating the 'heterosexual street': Lesbian productions of space. In N. Duncan (Ed.), *BodySpace: Destabilizing geographies of gender and sexuality* (pp. 146–155). London and New York: Routledge.

Vera-Gray, F. (2016). *Men's intrusion, women's embodiment: A critical analysis of street harassment*. Abingdon, Oxon: Routledge.

Vera-Gray, F. (2018). *The right amount of panic: How women trade freedom for safety*. Bristol: Polity Press.

Waitt, G. (2008). Urban festivals: Geographies of hype, helplessness and hope. *Geography Compass*, *2*(2), 513–537.

YouGov. (2018). *Harassment at festivals results*. YouGov. Retrieved June 21, 2018, from https://d25d2506sfb94s.cloudfront.net/cumulus_uploads/document/kuck5zispj/Press Association_180606_FestivalsHarrassment_w.pdf

7 Conceptualising safety and crime at UK music festivals
A gendered analysis

Hannah Bows, Hannah King and Fiona Measham[1]

Introduction

This chapter brings together conceptual developments across the social sciences to shed light on the underexplored subject of gender and safety at music festivals, a leisure location of growing interest to social scientists and of significant growth within the events industry. Drawing on debates in criminology, sociology, geography and gender studies, the authors consider the intersecting ways in which different social sciences conceptualise the gendering of social space and their applicability to contemporary UK music festivals, with a particular focus on sexual violence. Drawing on data from the UK's first study of gender, safety and crime at music festivals and contextualised in the wider literature on gender, crime, risk and licensed leisure space, the chapter raises questions regarding the distinctive features of commercialised music festivals, the extent to which they can be considered transgressive or countercultural spaces and what might be the distinctions, if any, of gendered sexual violence within such space. In doing so, this chapter explores how we might start to conceptualise festivals as gendered spaces from a criminological perspective and calls for a greater incorporation of criminological research within the broader festival studies field which can further the empirical and conceptual interrogation of music festivals.

Gender, space and leisure

A growing body of research has been undertaken on perceptions and experiences of 'space', particularly public spaces, and the ways in which social structures and constructs affect consumption and behaviours within them. For criminologists and sociologists, much of this work has focussed on examining the relationship between crime, 'deviance', subcultures and space and particularly violence between men in urban spaces, such as pubs, clubs and other licensed venues in late modern capitalist consumer society.

Relatedly, feminist scholars across geography and the social sciences for several decades have explored the gendered and sexualised nature of spaces, how they can be experienced differently by women and men and the ways in which these spaces may restrict or even exclude certain demographic groups.

In doing so, this work has highlighted the pervasiveness of sexual harassment and assault in public spaces, including, for example, public transport (UK Parliament, 2018), the street (Vera-Gray, 2016) and leisure spaces, such as sports clubs (Roper, 2016), gyms (Morris, 2019) and, of particular importance to this chapter, nightlife venues (e.g. Brooks, 2011; Drinkaware, 2015; Fileborn, 2016; Sheard, 2011; UK Parliament, 2018).

This latter body of work, situated at the borders of leisure studies, feminist and gender studies, criminology, sociology and geography, has been crucial in developing our understandings of the gendered and spatialised 'risks' that women experience across different public sites. In particular, this work has illuminated the role that commercialised heteronormative, sexualised spaces play in contributing to developing 'cultural atmospheres' where localised discrimination based on gender, ethnicity, social class, (dis)ability, sexual identity or other characteristics intersect and converge with broader Western cultures of "consumption, hedonism, risk, sex and heavy episodic alcohol use" (Kavanaugh, 2015, 242), which themselves shape and reinforce white heterosexual male power and privilege.

Within the discipline of leisure studies, feminist scholars have highlighted women's unequal access to leisure spaces and opportunities, enforced by patriarchal structures and gendered risks. This is true of most leisure spaces, including sport and outdoor leisure and tourist spaces, where women face higher physical risks of sexual harassment and assault (Kozak, 2007; Lepp & Gibson, 2003; Park & Reisinger, 2010; Yang et al., 2017) in these traditional masculinised and sexualised spaces (Jordan & Gibson, 2005; Wilson & Little, 2008).

Much of the early work documenting women's unequal access to, and restrictions within, different leisure spaces viewed space as static and women's experiences and reactions natural and automatic (Scraton & Watson, 1998). However, since the 1980s, there has been a shift to recognising not only the physicality of space but also the social, cultural and temporal features and constructs of different spaces, which produce gender inequality within different sites. Scraton and Watson (1998, 123) explained that the "universal explanations of leisure behaviours have been replaced by a concern to acknowledge differences, shifting focus to consumption and the spaces and places in which this occurs". Aitchison (1999) similarly noted that space was previously viewed as absolute and material but is now widely recognised as relative and symbolic, providing new ways of seeing and understanding leisure spaces. This marks a clear shift from seeing space as purely physical, to instead recognising it is the sociocultural and relative nature of space that is important.

Consequently, the synergies between gender relations and spatial relations began to be explored in academic research (Aitchison, 1999), building on feminist activism which began drawing attention to gender and space in the 1970s (e.g. Reclaim the Night marches, women's refuges). Early feminist geographers (e.g. Valentine, 1990; Pain, 1991) provided foundational research into the gendered nature of spaces, from their design to their function, which were built, constructed and maintained within the wider patriarchy. As Valentine (1990) noted, the spatial designs and segregation of various public spaces existed to ensure men's ability to exert control over women's use of space, but it is not purely the design of the space that is responsible. Rather, it is the social relations within particular spaces that enable the regulation and control of women's bodies. Similarly, the social and cultural structures which shape gendered perceptions, understandings and experiences of spaces are not static; they vary across time, space, place and between cultures and contexts (Massey, 1994). Green and Singleton (2006, 855) argued that the gendered nature of spaces and places both "reflects and has effects back on" the ways in which gender is constructed and understood. Moreover, these gendered constructions are sexed, classed and racialised, restricting access to women who do not conform to white, heterosexual middle-class ideals (Griffin et al., 2018).

One of the primary areas of academic interest has concerned women's fear in public spaces and, to a lesser extent, their experiences of sexual harassment and violence. Much of this research has drawn on national victimisation surveys or sources of quantitative data to estimate the fear of crime and actual experiences. This research has consistently shown that women have high levels of fear, particularly concerning sexual harassment and violence in public spaces, compared with men (for reviews and findings, see Pain, 1991; Stanko, 1995). The earlier work in this area focussed on public places such as parks and streets and was concerned with social and environmental factors which increased or decreased fear (e.g. lighting, wooded areas, presence of other people) (Jorgensen et al., 2012). Comparisons were frequently made between women's fear of sexual violence compared to their actual risk, reporting that women disproportionately feared these offences. However, as described earlier in this chapter, contemporary research has evidenced that sexual harassment and assault are frequent features of women's lives in both public and private spaces.

Violent attacks and incidents of sexual harassment can suggest to women that they are not meant to be in certain spaces at certain times and can have the effect of excluding women from various spaces or regulating their behaviour within them (Rose, 1993; Beebeejaun, 2017). It is well documented that women (and other marginalised or targeted groups) restrict and adjust their behaviour and movement through public spaces in order to manage this fear of encountering sexual violence in public spaces (Pain, 1991; Stanko, 1995; Vera-Gray, 2018). For example, women mentally map certain places in relation to their fear of male violence (Valentine, 1989). Much of this 'safety work' involves the hyper-vigilant woman making subtle behaviour adaptations involving the way certain places and times are negotiated (Vera-Gray, 2018; Valentine, 1989).

Pain (1997) described some of the 'precautions' women take to avoid violent victimisation, ranging from not answering the door to avoiding certain public places (particular streets or areas) to their choice of employment, leisure and social activities (Pain, 1997, 234). As Hollander (2001, 105) found,

> Women report constantly monitoring their environment for signs of danger, hesitating to venture outside alone or even in the company of other women,

asking men for protection, modifying their clothes ... and restricting their activities.... These strategies are simply part of daily life as a woman.

Thus it may be that the incidence of violence would be higher if it were not for the safety work women undertake as a result of their fear of crime in public spaces. Conversely, violence might be lower if greater numbers of women were visible in public spaces at all times of the day and night, 'reclaiming' the streets through collective presence.

Licensed leisure and crime

Criminologists have long been curious about the relationship between different spaces and crime. Since the 1920s, criminologists have been concerned with understanding not only who commits crime and who is victimised but also where crime occurs. The Chicago School, in particular the work of Shaw and McKay (1942), is often cited as one of the first major schools of thought to specifically examine the spatiality of criminal offending and victimisation. This early work was particularly interested in where offenders lived and socialised and identified criminogenic zones in urban areas of Chicago.

Through analysis of police-recorded crime data and victimisation surveys, a significant body of work has focussed on mapping the incidence of crime in order to examine causal effects and develop effective responses to reduce crime. Often described as ecological or situational crime research, this work has documented the socio-spatial elements and distribution of crime. It reveals that crime is not evenly distributed; patterns and concentrations of crime vary by crime type, but large urban areas have the highest levels of recorded violent and acquisitive crime (e.g. see ONS, 2019). However, it is not just the static space that is of concern to contemporary criminologists. Rather, it is the social organisation of spaces, and the sociocultural constructs within them, which provides environmental opportunities for crime and victimisation (Felson & Cohen, 1980).

In recent decades, both criminology and leisure studies have been interested in the relationship between, and incidence of, crime, 'deviance' and leisure spaces. As Smith and Raymen (2016, 63) note, "the study of 'leisure' is perhaps one of the central preoccupations of the social sciences". This has typically focussed on the problems associated with specific leisure spaces, such as public parks, skateboarding parks and nightlife venues, and has often been concerned with young people, the lower classes and the collective threat of association. In this context, leisure has been problematised, and the negative features and consequences of both the spaces and the activities within those spaces have been the focus of much of this work. As Measham (2004a, 2004b) has pointed out, the 'problem of leisure' has been a familiar and persistent concern in UK society back to Victorian times, particularly in relation to young people engaging in drinking, drug use and other behaviours commonly categorised as 'anti-social' or 'hooligan' (Pearson, 1983).

Unsurprisingly, night-time leisure has been a central concern in this literature, having grown rapidly over the last few decades to occupy a prominent, even dominant, position within the urban cityscape. In post-industrial Britain, the urban night-time economy (NTE) has been a legitimate focus of social and economic growth illustrated by the liberalisation of licensed leisure, the introduction of the notion of '24 hour' party cities in 1990s Britain and the normalisation of 'determined drunkenness' (Hadfield & Measham, 2009; Measham, 2004a, 2004b; Measham & Brain, 2005). By attracting visitors back to the city at night, these spaces were transformed into an evening economy, contributing to urban regeneration and economic growth (Van Liempt et al., 2015). These spaces are argued to form "an important part of identity for young consumers, characterised by a near universal adherence to intoxication and the suspension of the moral regulation and behavioural norms of the day-time" (Smith & Raymen, 2016, 69). Kavanaugh (2015) described NTE venues within the city downtown as premier destinations for young people looking to interact in the pursuit of hedonism and sexual courtship.

However, as Van Liempt et al. (2015) noted, this initial optimism about the benefits of the NTE has been displaced by growing concern about the commercialisation of these spaces, dominated by big branded names, the homogenisation of nightlife on offer, the exclusion of lower-class and non-white consumers and the environmental impacts of gentrification. Moreover, within these spaces, aggression and violence is common. In fact, studies have found that pubs and nightclubs are the most likely location for violent incidents between (primarily young) males (Leonard et al., 2002; Winlow & Hall, 2006). For Van Liempt et al. (2015, 408), the concentration of violence and anti-social behaviour in and around nightlife areas is not surprising, as these spaces are often "emotionally charged spaces offering many chances for the transgression of social norms that are taken for granted during the day". They provide the hotspots for violent crime and anti-social behaviour with recognised flashpoints created by overcrowding, intoxication and demand for public transport, communications and fast-food outlets (Tuck, 1989).

Thus the historical 'problem of leisure' came to be symbolised by the contemporary NTE with competing tensions from the loosening of licensing restrictions and rapid growth of the city-centre NTE fuelled by a 'revolution' in licensed leisure that included increasingly cheap alcohol, extended hours and a growing diversity of outlets, leading to growing concerns about young adult binge drinking in the early 2000s, at least in Northern Europe (Measham & Brain, 2005). Measham (2004a, 337) noted how "popular cultural worlds may be sanctioned and regulated whilst concurrently being problematised and criminalised". This echoes the early 'foundational' work of Chicago School Scholars that "targeted cabarets, taxi-dance halls, roundhouses and red-light districts as hotbeds of commercialised sex, gambling, bootlegging and organised crime" (Grazian, 2009, 908). Since then, there has been a growing interest in offending, victims and incidents of crime (Bottoms, 2007). Most recently, however, there has also been a shift in rhetoric, with nightlife in urban cities viewed as important cultural and socio-economic contributions (Grazian, 2009), combined with a recent fall in young adult alcohol consumption and a shift towards alternative pursuits during leisure time and in contemporary urban space (Conroy & Measham, 2019; Measham, 2008a).

Kavanaugh (2015, 240) argued,

> Emerging leisure economies are double-sided. While on one hand, they provide new opportunities for entrepreneurship and service sector employment, enable disposable spending, and so revitalize cities economically, on the other hand, they define leisure in ways that reproduce gender inequality, constrict individual behaviour to ensure market compliance and hedonistic forms of consumption, and so give rise to specific types of violence that they do little to effectively regulate or discourage.

Grazian (2009) shared these views, arguing that the romantic nostalgia held by urban ethnographers, such as Oldenberg (1989), who saw nightlife as a social leveller, rendering social inequalities temporarily irrelevant, is more of a dream than a reality. Grazian argued that, in the "neoliberal metropolis that characterises urban entertainment districts", gender differences and sexual harassment of women within these scenes is normalised (2009, 910). In fact, research confirms that gendered inequalities have been designed into the cultural and social structures of the NTE and are reinforced through the heteronormative practices of some venues operating within the NTE.

For Winlow and Hall (2006), Kavanaugh (2015) and other criminologists, the violence that occurs in nightlife contexts cannot be separated from broader societal structures. Indeed, Kavanaugh (ibid., 250) argued that there are "subterranean convergences" between the "value systems of individuals who participate in violence, the context where the violence plays out and the wider culture that permits such contexts to flourish". This wider (Western) culture is where risk, violence, casual sex and alcohol consumption are celebrated within certain social groups. Kavanaugh argued that the young, middle-class patron, the primary consumer of US neoliberal commercial nightlife, is seeking to "transcend the normative constraints of their mundane routines", but they become "constrained by another set of normative expectations: the intensely gendered regime of commercial nightclubs" (p. 252). In many ways, Kavanaugh is reflecting symbolic interactionist theories of gender and crime, which situate gender and sexuality "within the mundane activities of social life" (Jackson & Scott, 2010, 2). The following section will examine existing understandings of gender and sexuality in nightlife, particularly licensed leisure spaces.

Gender, sexuality and licensed leisure

Leisure and urban studies scholars, sociologists and feminist geographers have been concerned with the ways gender, sexuality and space are constructed, shaped and (re)produced in NTE spaces for the last three decades. Pilcher

(2011, 233) argued that, by examining women's "participation in sexualised leisure spaces, we can begin to analyse the intersection of geographies, gender, sexualities and space". As described earlier in this chapter, one of the key areas of interest for criminologists and leisure scholars researching the NTE has been explaining male-on-male violence and anti-social behaviour within the NTE. Although the role that alcohol plays in contributing to the incidence of violence has occupied much of the research, the relationships and intersections between gender, sexuality and masculinity have been central to developing understandings. The sexualised, hypermasculine and heteronormative structure of much of the NTE has provided a conceptual framework to understanding the highly gendered nature of these environments. It is a 'point of fact' that women and men have historically experienced much nightlife space as distinctly and overtly gendered (Grazian, 2009, 912). Thus, as others have noted (e.g. Nicholls, 2017), licensed leisure spaces within the NTE represent an interesting arena in which to explore some of the ways in which gender, space, safety and risk are inextricably linked.

Leisure spaces associated with the NTE, particularly drinking venues, such as bars, pubs and clubs, "have generally been highly gendered as masculine, with a limited range of 'respectable' drinking spaces and alcohol products aimed at women (Griffins et al., 2018, 186)". The consumption of alcohol, particularly excessive drinking, has "long operated as key markers of masculinity", whereas women's consumption of alcohol has traditionally been viewed as unfeminine (Griffin et al., 2018, 186; see also Hey, 1986; Measham, 2004b, 2008b, 2010). In the context of commercial nightlife venues, alcohol consumption has traditionally been "organised around a sexualised normative structure" (Kavanaugh, 2015, 248).

Although there are now larger numbers of women accessing increasingly diverse licensed spaces, with the new female drinker specifically targeted by the 1990s alcohol industry (Measham & Brain, 2005) and female clubbers as an integral part of the differently gendered nightlife space of raves and dance clubs (Henderson, 1993, 1997; Hunt et al., 2010; Hutton, 2004, 2006; Measham, 2004b; Measham et al., 2001), much of the NTE continues to be gendered and sexualised. This is evident in what Grazian (2009) describes as the feminisation and sex segregation of the nightlife industry. For example, in the British context, women continue to hold disproportionate numbers of lower paid and casual 'service' jobs within the industry, including bar work, waitressing and dancing/performing. Women may be part of the 'package' offered to male customers and venues draw upon the physical attractiveness of women and the 'sexual magnetism' (Grazian, 2009, 912) of female service staff who may be required to 'do gender' through sexualised femininity characteristic, such as tight and revealing clothing and eroticised behaviours. Thus bars, pubs and clubs can be hypersexualised spaces where women perform hyper-sexual forms of heteronormative, racialised and classed femininity expressed through a particular 'look' or behaviour for the benefit of the male consumer. Griffins et al. (2018) suggested that, in order to successfully negotiate these spaces, women must conform to a new form of hyper-sexual femininity in which they are independent but not feminist, get drunk with men but not 'like men' and look and act sexy for men's enjoyment but distance themselves from the 'drunken slut' (De Visser & McDonnell, 2011 cited in Griffins et al., 2018).

In other words, many women's experiences of nightlife spaces are still structured by assumptions about their (hetero)sexual availability (Sheard, 2011) and they remain subject to the male gaze, sexual harassment and assault. For example, security staff at nightclubs and bars have been found to contribute to the culture of masculinity and sexism which is often present in bars (Hobbs et al., 2003; Tomsen, 1997; Winlow & Hall, 2006). Consequently, Nicholls (2017) argued that the heteronormativity of nightlife spaces continues to act as a form of governance to police and control women, and the gender differentiation within the NTE continues to define and preserve the culture of these urban spaces. Similarly, Brooks (2011, 334) has argued "Issues surrounding women's access to leisure activities and their use of public space are heightened when they seek to socialise in bars, pubs, and clubs; women who enter these spaces are subject to heightened (wanted or unwanted) male attention".

Licensed leisure spaces are frequently associated with sexual harassment and violence against women and to a lesser extent, men, particularly those linked to the NTE, which are situated within the 'sexualised city' (Hubbard & Colosi, 2015). In the UK, a Drinkaware study (2015) reported 54% of women and 15% of men aged 18–24 experience sexual harassment during a night out. Mellgren et al. (2018) found that sexual harassment of Swedish women occurred frequently at clubs and restaurants, and, although many of these incidents would not meet the legal definitions of a crime, the regularity of these low-level incidents contribute to what Kavanaugh (2015) described as a 'cultural atmosphere' where unwanted sexual attention becomes accepted as a normal part of being in public places. For Kavanaugh (2015), the culture of commercial nightlife further reinforces the sentiments of hypermasculine performance and heterosexual power amongst patrons, creating a space that facilitates and legitimises harassment and degradation of women.

This can result in leisure spaces, particularly alcohol venues in the NTE, being viewed as 'risky spaces' by women. As Green and Singleton (2006, 854) note, leisure is a key arena for risk-taking behaviour, and it is deeply gendered, both in terms of the spaces and places that young women occupy and their behaviour within such spaces. Such behaviours are also overlaid by differences of age, class, sexuality, 'race', ethnicity, (dis)ability and culture, although some of these differences are less well acknowledged than others (i.e. disability discrimination is only just starting to be recognised and responded to in the NTE and is significantly behind other areas on inequality). Unsurprisingly then, leisure spaces, as with other public spaces, are perceived in terms of their safety and possible threat of male violence (Scraton & Watson, 1998). Risk-taking (e.g. excessive alcohol consumption, substance misuse, violence and anti-social behaviour) is generally encouraged amongst male consumers and viewed as a marker of masculinity, whereas women may be expected to avoid engagement with risky places and behaviours. As Yang et al. (2017, 90) point out, in general, women's risk-taking behaviour is more likely to be negatively evaluated compared to men's because risk-taking is associated with the construction of masculinity, whereas risk aversion is a desirable trait of femininity (Elsrud, 2001; Campbell, 2005; Laurendeau, 2008; Olstead, 2011).

Scraton and Watson (1998) further problematise the concept of risk. They argue that risk is often positioned as an active choice and part of the excitement and pleasure of leisure consumption, but this is a heteronormative, masculine understanding that obscures the reality of 'risk' for female consumers. They argue that, for women, risk is not a choice but rather an inherent feature of leisure environments, which must be constantly assessed and managed. It is not an exciting part of the experience but rather an unavoidable element. Regardless of the increase in women's involvement in the NTE, there is a prevailing assumption that by occupying these spaces women are inherently vulnerable to male violence and have unnecessarily placed themselves at risk (de Crespigny, 2001).

It is against this backdrop of gendered licensed leisure space dating back to Victorian England that alternative, counterculture and transgressive leisure space has developed, including raves and dance clubs since the late 1980s (Thornton, 1995) and music festivals since the 1960s (Clarke, 1982) to which we now turn our attention.

Conceptualising gender and crime at UK music festivals

A festival can be loosely defined as an organised series of events typically in one location and with a unifying theme. Such definitions belie the vast range of different types of festivals; however, with at least 17 different 'types' of festivals identified by Stone (2009), varying in size, style and patron demographics. Some of this range has been explored within the multidisciplinary field of festival studies – the empirical, theoretical and policy analysis of festivals – that developed over the last two decades to explore a wide range of festival spaces. Studies on, for example, art festivals (Quinn, 2005; Waterman,1998), folk festivals (Quinn & Wilks, 2017), local community festivals (Clarke & Jepson, 2011) and food and wine festivals (Yuan et al., 2008), have developed our understanding of the growing diversity and popularity of festivals and their significance in terms of social relations, economic investment and community cohesion. Whilst there are criminological contributions considering issues of transgression, hedonism and intoxication, discussed earlier, the gendering of festival space and issues of safety and violence within festivals have not been a primary concern to those working or researching this space (Gisbert and Rius-Ulldemolins, 2019).

The authors have turned their attention to this issue with data collection at UK music festivals. Music festivals may be held in tents erected in large fields where patrons camp over the course of several days (often referred to as green-field festivals) or can be city festivals on one or more days with or without camping and located in local parks, existing buildings or in city streets. The disparities in defining and counting such festivals is evident in that UK Music (2017) estimates there were around 230 festivals in 2016 attended by more than 4 million people; whereas CGA (2019) estimates that there were 700 UK music festivals attended by 7.1 million customers in 2018, and Mintel (2018) estimates that there were 918 UK festivals in 2018, more than double that of a decade earlier.

Regardless of their composition, music festivals, as with much of the music, events and hospitality industries, are widely acknowledged as heavily gendered/male-dominated businesses, although women's participation, as artists and as production staff, has increased significantly over the last decade. For example, in 2016, 60% of UK music festival visitors were female, an increase from 37% in 2015 (Statista, 2016). This gender shift mirrors changes across the broader licensed leisure industry.

Whilst music festivals may share some characteristics with other licensed leisure spaces, there are also significant differences, since music "festivals are unique in their size, location and layout: are held at both day and night-time, are relatively infrequent, of long duration and large crowd size" (Dilkes-Frayne, 2016, 1). Since music festivals first emerged in their current form in the late 1960s (Clarke, 1982), their core appeal to their young adult customers – selling images of liminality, temporary freedom and 'time out' from everyday life (Pielichaty, 2015) – have led music festivals to be associated with crime, 'deviance' and transgression, at least to the wider public. To the onlooker, transgressive festival behaviours can range from increased drinking, drunkenness, drug use and nudity, through to anti-social behaviour and violence, as well as countercultural rebellion or resistance, a carnivalesque inversion of social norms and a 'Peace, Love, Unity, Respect' ethos (shared more recently associated with the rave scene) that reached mythic status in festival folklore, such as with the Woodstock festival in the US in 1969.

As music festivals have expanded both in numbers and capacity and given their conceptualisation as sites of transgression, so the gendering of music festivals and the social construction and performance of gender within them are emerging as areas for fertile academic interest (e.g. Bhardwa, 2013; Motl, 2018; Pernecky et al., 2019). Tokofsky (1999 cited in Pielichaty, 2015, 239) suggested that "gender features centrally in festival environments because of a perceived opportunity for freedom and liberation; therefore, the chance for festival goers to play at the edges of gender seems obvious". Green (1998) concurred that music festivals can be viewed as sites for 'gender work' where masculinities and femininities are constructed and produced.

Within criminology and sociology, in the last decade, researchers have turned their attention to large music festivals and dance club tourism as sites of transgression, liminality and intoxication (e.g. Bhardwa, 2013; Hesse & Tutenges, 2008; Morey et al., 2014; Ruane, 2017; Turner, 2018). Within this emergent field of festival studies, however, studies have rarely extended to an examination of gendered risks and experiences at music festivals, and this has been undertaken predominantly by students (e.g. Motl, 2018; Pernecky et al., 2019; Pielichaty, 2015). This is in stark contrast to coverage of music festivals in the online and print media, which devotes significant space to these events each year, including reporting on incidences of sexual harassment, violence and crime more broadly.

Studying gender, safety and crime at UK music festivals: setting a research agenda

It is clear that music festivals, like other leisure spaces, are gendered spaces and the experiences of music festival-goers in relation to crime and safety is differentially gendered. However, given that the promise of liminality, transgression and 'freedom' is a distinctive characteristic of music festivals (illustrated in festival names, such as *Secret Garden Party*, *Lost Village* and *Wilderness*), we anticipate additional complexities in unpicking the role of gender in festival crime and, therefore, additional value in understanding sexual violence in festival space.

We recently undertook the first UK study to examine safety and crime at music festivals, including specifically focussing on the gendered differences between men and women who attended a festival in the previous 12 months. The main study surveyed 450 self-selecting festival-goers online about their perceptions of safety and experiences of crime at UK music festivals. This included 285 women respondents. The findings from their responses revealed that most women feel relatively safe at festivals. Respondents reported feeling usually safe (50.9%) or always safe (35.1%). Very few respondents felt rarely (2.8%) or never (0.4%) safe. A recent study of perceptions of safety and sexual violence at Australian music festivals similarly found that the vast majority of participants (men and women combined) reported that they either 'usually' (61.5%) or 'always' (29%) feel safe at music festivals (Fileborn et al., 2018), See Chapter 6 for more information. However, despite female respondents reporting an overall feeling of safety at music festivals, various personal, social and environmental features increased or reduced their feelings of safety along with some spaces within festivals that they felt particularly unsafe. Furthermore, the two key concerns of women regarding crime and safety on-site were sexual harassment and sexual assault. Overall, a third of women reported experiencing sexual harassment and 8% sexual assault at a UK festival in the previous 12 months.

This resonates with feminist geographers' analyses of gender and space and studies on gender and fear of crime discussed earlier. A YouGov (2018) study highlighted the prevalence of sexual harassment at festivals in the UK, with 43% of female festival-goers under the age of 40 experiencing unwanted sexual behaviour. However, this is a vastly lower level than that reported in a recent US survey, where over 90% of women said they had been sexually harassed at a music festival or music gig/venue (OMMB, 2017). The picture is yet more complex when comparing to figures for other leisure spaces, such as bars and clubs, where a Drinkaware (2015) survey found that 54% of women aged 18–24 experience sexual harassment on a night out. Whilst sexual violence at music festivals hits the headlines every summer, academic research on the issue is scarce, and there remain major gaps in knowledge.

Historically, festivals have been understood as liminal spaces providing temporary freedom and 'time out' from everyday life (Bhardwa, 2013; Pielichaty, 2015). As a result, and mostly through sensational media coverage, they have become associated with crime, deviance and transgression, in particular excessive drinking and drug taking. Hayward (2002) notes how transgressive leisure and carnivalesque pleasure offer an opportunity for escapism not just from boredom but also from the insecurity and 'hyper-banalisation' of everyday life in which people feel increasingly over-controlled, not only by agents of the state but also in a cultural and economic sense. However, the extent to which festivals are genuinely radical and countercultural is questionable, given the commodification of transgression and the intense control of contemporary festivals (Haydock, 2015). According to Turner and Measham (2019), festivals represent carnivalesque realms that merge opportunities for transgressive pleasure with elements of risk, danger and subversion. Within this ethereal context, consumed with a sense of ambivalent well-being, their participants' perceptions of risk were distorted, leading to changed behaviour, most notably patterns of increased alcohol and drug use that were atypical of their consumption in everyday life.

This raises a number of questions for the emergent field of festival studies. Given the mass commodification of festivals, often into Instagrammable 'brandfests', to what extent can music festivals (still) really conceptualised as 'wild zones'? Are festivals safe spaces to engage in risky behaviour? What is the relationship between crime, leisure and transgression within music festivals? What are the risks and what is the crime picture within these spaces? How does all of this impact festival-goers' perceptions of crime and safety? If, as Turner and Measham (2019) have found, festival-goers have altered perceptions, behaviours and assessments of risk, how does this impact intimate relationships and sexual violence? If festivals are an area of atypical behaviour compared to everyday life, but are still gendered, how does this impact people's experiences of sexual violence? How do women 'do gender' at festivals? If festivals are 'wild zones', do women alter their behaviour (their 'safety work') in different ways to other spaces (Vera-Gray, 2018)? How do men view women within festivals, and how is all of this reflected in the responses of on-site services? Along with their endurance as a site of pleasure and pilgrimage for millions of people every summer, this makes music festivals an interesting and important arena to study further.

Conclusion and future directions

This chapter has considered what is currently known about gender, leisure space and violent crime and how existing conceptual understandings of gender and space, particularly licensed leisure spaces, may provide useful starting points for further investigation of gender and crime at music festivals. We have noted that despite the significant body of research on gender, violence, licensed leisure and the NTE, and despite music festivals occupying an increasingly prominent position in young adult social calendars, there is comparatively little criminological research within the field of festival studies on large-scale commercialised music festivals and patterns and prevalence of violent crime within them. Drawing on existing work in the social sciences and the emerging findings from a study we conducted examining safety and crime at UK music festivals, we have argued that the distinct spatial, temporal and cultural features of music festival sites require specific empirical and conceptual consideration; existing understandings of leisure spaces and gender may provide useful wider contexts, but the findings from these fields do not necessarily translate to festival spaces. We suggest instead that criminological research on gender and sexual violence is incorporated within the broader field of festival studies to further the empirical and conceptual interrogation of music festivals and that it includes an intersectional analysis of consumption, risk and experiences within these spaces, incorporating not only gender but also race, class, sexuality and (dis)ability in order to develop comprehensive understandings of sexual violence and harassment within festival spaces.

Note

1 Authors are in alphabetical order.

References

Aitchison, C. (1999). New cultural geographies: The spatiality of leisure, gender and sexuality. *Leisure Studies*, *18*(1), 19–39.

Beebeejaun, Y. (2017). Gender, urban space, and the right to everyday life. *Journal of Urban Affairs*, *39*(3), 323–334.

Bhardwa, B. (2013). Alone, Asian and female: The unspoken challenges of conducting fieldwork in dance settings. *Dancecult: Journal of Dance Music Culture*, *5*(1), 39–60.

Bottoms, A. E. (2007). Place, Space, Crime and Disorder. In M. Maguire, R. Morgan & R. Reiner (Eds.), *Oxford handbook of criminology*, 4th edn. Oxford: Oxford University Press.

Brooks, O. (2011). "Guys! Stop doing it!": Young women's adoption and rejection of safety advice when socializing in bars, pubs and clubs. *The British Journal of Criminology*, *51*(4), 635–651.

Campbell, A. (2005). Keeping the "lady" safe: The regulation of femininity through crime prevention. *Critical Criminology*, *13*(2), 119–140.

CGA. (2019). *Your future in festivals: How to stand out from the crowd*. Retrieved from www.cga.co.uk/wp-content/uploads/2019/05/YourFutureinFestivals-CGA-Insights.pdf

Clarke, A., & Jepson, A. (2011). Power and hegemony within a community festival. *International Journal of Festival and Event Management*, *2*(1), 7–19.

Clarke, M. (1982). *The politics of pop festivals*. London: Junction Books.

Conroy, D., & Measham, F. (Eds.). (2019). *Young adult drinking styles: Current perspectives on research, policy and practice*. Switzerland: Springer Nature.

de Crespigny C. (2001). Young women, pubs and safety in Alcohol, Young Persons and Violence, Chapter: Young women, pubs and safety. In Alcohol, Young Persons and

Violence. In P. Williams (Ed.), *Australian Institute of Criminology, Canberra, ACT*. Australian Institute of Criminology Research and Public Policy Series No. 35.

De Visser, R., & McDonnell, L. (2011). Masculine capital and men's health-related behaviour: Abstract ID:# 0015. *Journal of Men's Health, 8*(3), 234–234.

Dilkes-Frayne, E. (2016). Drugs at the campsite: Socio-spatial relations and drug use at music festivals. *International Journal of Drug Policy, 33*, 27–35.

Drinkaware. (2015). *Sexual harassment tops list of risks for female students on nights out*. Retrieved September 7, 2019, from www.drinkaware.co.uk/press/sexual-harassment-tops-list-of-risks-for-female-students-on-nights-out/

Elsrud, T. (2001). Risk creation in traveling backpacker adventure narration. *Annals of Tourism Research, 28*(3), 597–617.

Felson, M., & Cohen, L. E. (1980). Human ecology and crime: A routine activity approach. *Human Ecology, 8*(4), 389–406.

Fileborn, B., Wadds, P., & Tomsen, S. (2018). *Safety, sexual harassment and assault at Australian music festivals-final report*. Sydney: University of New South Wales.

Gisbert, V., & Rius-Ulldemolins, J. (2019). Women's bodies in festivity spaces: Feminist resistance to gender violence at traditional celebrations. *Social Identities*, online first.

Grazian, D. (2009). Urban nightlife, social capital, and the public life of cities. *Sociological Forum, 24*(4), 908–917.

Green, E. (1998). Women doing friendship: An analysis of women's leisure as a site of identity construction, empowerment and resistance. *Leisure Studies, 17*(3), 171–185.

Green, E., & Singleton, C. (2006). Risky bodies at leisure: Young women negotiating space and place. *Sociology, 40*(5), 853–871.

Griffin, C., Bengry-Howell, A., Riley, S., Morey, Y., & Szmigin, I. (2018). "We achieve the impossible": Discourses of freedom and escape at music festivals and free parties. *Journal of Consumer Culture, 18*(4), 477–496.

Hadfield, P., & Measham, F. (2009). A review of nightlife and crime in England and Wales. In P. Hadfield (Ed.), *Nightlife and crime* (pp. 17–48). Oxford: Oxford University Press.

Haydock, W. (2015). Understanding English alcohol policy as a neoliberal condemnation of the carnivalesque. *Drugs: Education, Prevention and Policy, 22*(2), 143–149.

Hayward, K. (2002). The vilification and pleasures of youthful transgression. In J. Muncie, G. Hughes, & E. McLaughlin (Eds.), *Youth justice: Critical readings*. London: Sage Publications.

Henderson, S. (1993). Luvdup and de-elited: Responses to drug use in the second decade. In P. Aggleton, P. Davies, & G. Hart (Eds.), *AIDS: Facing the second decade* (pp. 119–130). London: Falmer.

Henderson, S. (1997). *Ecstasy: Case unsolved*. London: Pandora.

Hesse, M., & Tutenges, S. (2008). Music and substance preferences among festival attendants. *Drugs and Alcohol Today, 12*(2), 82–88.

Hey, V. (1986). *Patriarchy and pub culture*. London: Tavistock.

Hobbs, D., Hadfield, P., Lister, S., & Winlow, S. (2003). *Bouncers: Violence and governance in the night-time economy*. Oxford: Oxford University Press.

Hollander, J. A. (2001). Vulnerability and dangerousness: The construction of gender through conversation about violence. *Gender & Society, 15*(1), 83–109.

Hubbard, P., & Colosi, R. (2015). Taking back the night? Gender and the contestation of sexual entertainment in England and Wales. *Urban Studies, 52*(3), 589–605.

Hunt, G., Moloney, M., & Evans, K. (2010). *Youth, drugs, and nightlife*. London: Routledge.

Hutton, F. (2004). Up for it, mad for it? Women, drug use and participation in club scenes. *Journal of Health, Risk & Society*, *6*(3), 223–237.

Hutton, F. (2006). *Risky pleasures: Club cultures and feminine identities*. Aldershot: Ashgate.

Jackson, S., & Scott, S. (2010). *Theorizing sexuality*. Maidenhead: Open University Press.

Jordan, F., & Gibson, H. (2005). We're not stupid. But we'll not stay home either: Experiences of solo women travelers. *Tourism Review International*, *9*(2), 195–211.

Jorgensen, L. J., Ellis, G. D., & Ruddell, E. (2012). Fear perceptions in public parks: Interactions of environmental concealment, the presence of people recreating, and gender. *Environ Behavior*, *45*(7), 803–820.

Kavanaugh, P. R. (2015). The social organization of masculine violence in nighttime leisure scenes. *Criminal Justice Studies*, *28*(3), 239–256.

Kozak, M. (2007). Tourist harassment: A marketing perspective. *Annals of Tourism Research*, *34*(2), 384–399.

Laurendeau, J. (2008). Gendered risk regimes: A theoretical consideration of edgework and gender. *Sociology of Sport Journal*, *25*(2008), 293–309.

Leonard, K. E., Quigley, B. M., & Collins, L. R. (2002). Physical aggression in the lives of young adults: Prevalence, location, and severity among college and community samples. *Journal of Interpersonal Violence*, *17*(5), 533–550.

Lepp, A., & Gibson, H. (2003). Tourist roles, perceived risk and international tourism. *Annals of tourism research*, *30*(3), 606–624.

Massey, D. (1994). *Space, place and gender*. Oxford: Polity Press.

Measham, F. (2004a). The decline of ecstasy, the rise of "binge" drinking and the persistence of pleasure. *Probation Journal*, *51*(4), 309–326.

Measham, F. (2004b). Play space: Historical and socio-cultural reflections on drugs, licensed leisure locations, commercialisation and control. *International Journal of Drug Policy*, *15*(5–6), 337–345.

Measham, F. (2008a). The turning tides of intoxication: Young people's drinking in Britain in the 2000s. *Health Education*, *108*(3), 207–222.

Measham, F. (2008b). A history of intoxication: Changing attitudes to drunkenness and excess in the United Kingdom. In M. Martinic & F. Measham (Eds.), *Swimming with crocodiles: The culture of extreme drinking* (pp. 13–36), ICAP Series on Alcohol in Society, Volume 9. New York and Abingdon: Routledge.

Measham, F. (2010). Drunkenness: A historical and contemporary cross-cultural perspective: "A voluntary madness". In A. Fox & M. MacAvoy (Eds.), *Expressions of drunkenness (Four hundred rabbits)* (pp. 121–153), ICAP Series on Alcohol in Society, Volume 11. New York: Routledge.

Measham, F., Aldridge, J., & Parker, H. (2001). *Dancing on drugs: Risk, health and hedonism in the British club scene*. London: Free Association Books.

Measham, F., & Brain, K. (2005). "Binge" drinking, British alcohol policy and the new culture of intoxication. *Crime, Media, Culture: An International Journal*, *1*(3), 263–284.

Mellgren, C., Andersson, M., & Ivert, A. K. (2018). "It happens all the time": Women's experiences and normalization of sexual harassment in public space. *Women & Criminal Justice*, *28*(4), 262–281.

Mintel. (2018). *Music concerts and festivals: UK – August 2018*. London: Mintel Group Ltd.

Morey, Y., Bengry-Howell, A., Griffin, C., Szmigin, I., & Riley, S. (2014). Festivals 2.0: Consuming, producing and participating in the extended festival experience. In *The festivalization of culture: Celebration, identity and politics* (pp. 251–269). Farnham, Surrey: Ashgate Publishing.

Morris, N. (2019). Women are quitting their gym memberships because of sexual harassment. *The Metro*, June 20. Retrieved from https://metro.co.uk/2019/06/20/women-quitting-gym-memberships-sexual-harassment-10016822/?ito=cbshare

Motl, K. (2018). *"Well, don't walk around naked . . . unless you're a girl": Gender, sexuality, and risk in Jamtronica festival subcultural scenes*. University of Kentucky, Theses and Dissertations – Sociology, 38. Retrieved from https://uknowledge.uky.edu/sociology_etds/38https://doi.org/10.13023/etd.2018.332

Nicholls, E. (2017). "Dulling it down a bit": Managing visibility, sexualities and risk in the Night Time Economy in Newcastle, UK. *Gender, Place & Culture*, *24*(2), 260–273.

Office for National Statistics. (2019). *Crime in England and Wales: Year ending March 2019*. London: ONS. Retrieved September 7, 2019, from www.ons.gov.uk/peoplepopulationandcommunity/crimeandjustice/bulletins/crimeinenglandandwales/yearendingmarch2019

Oldenburg, R. (1999). *The great good place: Cafes, coffee shops, bookstores, bars, hair salons, and other hangouts at the heart of a community*. Boston: Da Capo Press.

Olstead, R. (2011). Gender, space and fear: A study of women's edgework. *Emotion, Space and Society*, *4*(2), 86–94.

Our Music My Body. (2017). *2017 campaign and survey report*. Retrieved from www.ourresilience.org/wp-content/uploads/2018/03/OMMB-2017-Campaign-Survey-Report.pdf

Pain, R. H. (1991). Space, sexual violence and social control: Integrating geographical and feminist analyses of women's fear of crime. *Progress in Human Geography*, *15*(4), 415–431.

Pain, R. H. (1997). Social geographies of women's fear of crime. *Transactions of the Institute of British Geographers*, *22*(2), 231–244.

Park, K., & Reisinger, Y. (2010). Differences in the perceived influence of natural disasters and travel risk on international travel. *Tourism Geographies*, *12*(1), 1–24.

Pearson, G. (1983). *Hooligans: A history of respectable fears*. London: MacMillan.

Pernecky, T., Abdat, S., Brostroem, B., Mikaere, D., & Paovale, H. (2019, in press). Sexual harassment at festivals and events: A student perspective. *Event Management*.

Pielichaty, H. (2015). Festival space: Gender, liminality and the carnivalesque. *International Journal of Event and Festival Management*, *6*(3), 235–250.

Pilcher, K. E. M. (2011). A "sexy space" for women? Heterosexual women's experiences of a male strip show venue. *Leisure Studies*, *30*(2), 217–235.

Quinn, B. (2005). Arts festivals and the city. *Urban Studies*, *42*(5–6), 927–943.

Quinn, B., & Wilks, L. (2017). Festival heterotopias: Spatial and temporal transformations in two small-scale settlements. *Journal of Rural Studies*, *53*, 35–44.

Roper, E. A. (2016). Concerns for personal safety among female recreational runners. *Women in Sport and Physical Activity Journal*, *24*(2), 91–98.

Rose, G. (1993). *Feminism and geography: The limits of geographical knowledge*. Cambridge, UK: Polity Press.

Ruane, D. (2017). "Wearing down of the self": Embodiment, writing and disruptions of identity in transformational festival fieldwork. *Methodological Innovations*, *10*(1), 1–11.

Scraton, S., & Watson, B. (1998). Gendered cities: Women and public leisure space in the "postmodern city". *Leisure Studies*, *17*(2), 123–137.

Shaw, C. R., & McKay, H. D. (1942). *Juvenile delinquency and urban areas: A study of rates of delinquents in relation to differential characteristics of local communities in American cities*. Chicago: University of Chicago Press.

Sheard, L. (2011). "Anything could have happened": Women, the night-time economy, alcohol and drink spiking. *Sociology, 45*(4), 619–633.

Smith, O., & Raymen, T. (2016). Deviant leisure: A criminological perspective. *Theoretical Criminology, 22*(1), 63–82.

Stanko, E. (1995). *Everyday violence: How men and women experience sexual and physical danger*. London: Pandora.

Statista. (2016). *Gender profile of festival-goers in the United Kingdom (UK) from 2012 to 2016*. Retrieved September 7, 2019, from www.statista.com/statistics/282836/gender-distribution-of-visitors-to-uk-music-festivals/

Stone, C. (2009). The British pop music festival phenomenon. In J. Ali-Knight, M. Roberston, A. Fyall, & A. Ladkin (Eds.), *International perspectives of festivals and events: Paradigms of analysis* (pp. 205–224). London: Elsevier.

Thornton, S. (1995). *Club cultures: Music, media and subcultural capital*. Cambridge: Polity Press.

Tokofsky, P. (1999). Masking gender: A German carnival custom in its social context. *Western Folklore, 58*(3/4), 299–318.

Tomsen, S. (1997). A top night: Social protest, masculinity and the culture of drinking violence. *British Journal of Criminology, 37*(1), 90–102.

Tuck, M. (1989). *Drinking and disorder: A study of non-metropolitan violence*, Home Office Research Study No. 108. London: HMSO.

Turner, T. (2018). Disneyization: A framework for understanding illicit drug use in bounded play spaces. *International Journal of Drug Policy, 58*, 37–45.

Turner, T., & Measham, F. (2019). Into the woods: Contextualising atypical intoxication by young adults in music festivals and nightlife tourist resorts. In D. Conroy & F. Measham (Eds.), *Young adult drinking styles: Current perspectives on research, policy and practice*. London: Palgrave Macmillan.

UK Music. (2017). *UK live music attendance and music tourism in 2017*. Retrieved from www.ukmusic.org/assets/general/Live_Music.pdf

UK Parliament. (2018). *Women and girls' safety on public transport*. Retrieved September 7, 2019, from https://publications.parliament.uk/pa/cm201719/cmselect/cmwomeq/701/70108.htm

Valentine, G. (1989). The geography of women's fear. *Area, 21*(4), 385–390.

Valentine, G. (1990). Women's fear and the design of public space. *Built Environment, 16*(4), 288–303.

Van Liempt, I., Van Aalst, I., & Schwanen, T. (2015). Introduction: Geographies of the urban night. *Urban Studies, 52*(3), 407–421.

Vera-Gray, F. (2016). *Men's intrusion, women's embodiment: A critical analysis of street harassment*. London: Routledge.

Vera-Gray, F. (2018). *The right amount of panic: How women trade freedom for safety in public*. Bristol: Policy Press.

Waterman, S. (1998). Carnivals for elites? The cultural politics of arts festivals. *Progress in Human Geography, 22*(1), 54–74.

Wilson, E., & Little, D. E. (2008). The solo female travel experience: Exploring the 'geography of women's fear'. *Current Issues in Tourism, 11*(2), 167–186.

Winlow, S., & Hall, S. (2006). *Violent night: Urban leisure and contemporary culture*. Oxford: Berg.

Yang, E. C. L., Khoo-Lattimore, C., & Arcodia, C. (2017). A systematic literature review of risk and gender research in tourism. *Tourism Management, 58*, 89–100.

YouGov (2018). *Two in five young female festival goers have been subjected to unwanted sexual behavior [online]*. Retrieved from https://yougov.co.uk/topics/lifestyle/articles-reports/2018/06/21/two-five-young-female-festival-goers-have-been-sub. Last accessed 4th February 2020.

Yuan, J., Morrison, A., Cai, L., & Linton, S. (2008). A model of wine tourist behaviour: A festival approach. *International Journal of Tourism Research, 10*, 207–219.

8 Gender-based violence amongst music festival employees

Cassandra Jones

Introduction

New feminist movements (e.g. #MeToo) have brought to the forefront of public consciousness the widespread prevalence of workplace gender-based violence (WGBV) and the scale of the impact on women's everyday lives. These movements have yet to influence the workplace environment of music festivals in the UK, which may be due in part to the academic and industry focus on audiences as consumers (e.g. Szmigin et al., 2017; Wilks, 2011) who characterise festivals as facilitating experiences free from the constraints of everyday lives (e.g. Griffin et al., 2018; Morgan, 2007). While it is important – and often contested – to understand what festivals can create for audiences, the experiences of festival employees, who construct these so-called liberating spaces for audiences, have been marginalised. This is no more apparent than when it comes to acknowledging and addressing sexual harassment. High-profile artists (e.g. Chapple, 2017), media (e.g. Long, 2018) and research (e.g. Fileborn et al., 2018) have fostered awareness of sexual harassment experienced by audiences, demonstrating that festivals are no different than other social sites and in need of cultural change. However, full and long-lasting cultural change requires addressing everyone in music festival sites, including employees. Understanding employees' experiences is a first step to creating pathways of cultural change, which is becoming more and more of an imperative for two reasons. One, festivals are an integral component in live music, one of the fastest growing creative industries in the UK (Chapple, 2018; UK Music, 2018), and two, the Equalities and Human Rights Commission (2018) and the Women and Equalities Committee (2018) called for all workplaces to take action to prevent and respond to WGBV.

This chapter answers these calls by bringing together decades of research on gender-based violence (GBV) in workplaces with more recent research on music festivals. A general definition of GBV is set out first in order to provide the larger societal context in which WGBV occurs. The focus then narrows to sexual harassment, one of the most commonly researched forms of WGBV, with a brief review of research conducted in workplaces with traditional hierarchical power structures. Next, research on workplace bullying is laid out, followed by a review of the debate on the links between sexual harassment and bullying to encourage the reader not to silo forms of WGBV from each other. To date, no research on sexual harassment and bullying amongst music festival employees can be located, so the power structures and characteristics of music festival workplaces are explored first to provide the context to apply knowledge developed from research in other workplaces. This sets the stage to propose how sexual harassment and bullying may manifest in music festival workplaces and how they may have an impact on employees. The chapter ends with research and policy recommendations that may aid music festivals in answering the calls of the Equalities and Human Rights Commission (2018) and the Women and Equalities Committee (2018).

WGBV

The Istanbul Convention defined GBV as any act of violence and abuse that disproportionately affects women and is rooted in systematic power differences and inequalities between men and women (Hester & Lilley, 2014). While more women than men are victim-survivors, men may also experience GBV (Hester et al., 2017; Jones, 2019). Individuals of any socio-demographic background may experience GBV, though individuals occupying positions of less social power (e.g. women, BME, disabled) have a greater chance of experiencing it (e.g. Coulter & Rankin, 2017). The World Health Organisation (2013, 5) found those experiencing GBV were more than two times as likely to experience mental health issues and thus declared it "a global health problem of epidemic proportions", making it a pressing social issue to understand and prevent in every social site, including the workplace.

Men are the primary perpetrators of GBV (Black et al., 2011; Garcia et al., 2006; Stöckl et al., 2013). Anyone including (but not limited to) family members, coworkers, parents, spouses and neighbours may use violent and abusive behaviours (Watts & Zimmerman, 2002), which can be manifested in a variety of forms (e.g. cyber violence, female genital mutilation, intimate partner violence). One of the most detrimental and widely researched forms in the UK is sexual violence, which encompasses a spectrum of behaviours (Kelly, 1987) from rape to assault to sexual harassment. Research on WGBV in the UK was conducted mostly within the framework of sexual harassment or bullying. The term 'sexual harassment' developed from the work of MacKinnon (1979) and Farley (1978), who named and challenged behaviours that were accepted as a 'normal' part of women's workplace experiences. A substantial body of international academic, legal and policy research has developed since (e.g. Jagsi et al., 2016; Klatt, 2018; McDonald, 2012; Tinkler et al., 2015), but an overarching problem has been and continues to be inconsistent conceptual and operational definitions, which is in large part due to the lack of consensus on 1) the specific behaviours constituting sexual harassment, 2) if only victim-survivors must experience consequences or if bystanders and co-workers can be affected as well and 3) if sexism is only one type of sexual harassment (Pina et al.,

2009). Further conceptual complications arose with the emergence of research on bullying in the 1990s when the name used to describe these behaviours varied (e.g. 'mobbing' in Scandinavia, 'bullying' in the UK, 'psychological aggression' and 'workplace incivility' in North America), and the term bullying was often used interchangeably with sexual harassment (Jones, 2006). Some (e.g. Vartia & Hyyti, 2002) proposed sexual harassment is a type of bullying which in turn led to debates on the relationship between sexual harassment and bullying (Jones, 2006). The debate is reviewed first before going on to discuss WGBV amongst music festival employees.

Sexual harassment and bullying: never the twain shall meet?

The debates on the relationship between sexual harassment and bullying are grounded in the theoretical frameworks and definitions used in interdisciplinary studies. Within the field of sexual harassment, scholars have used a diversity of theoretical perspectives, with natural/biological theories on one side of the spectrum that suggest sexually harassing behaviours are a result of men's inherent sexual nature towards women and on the other side of the spectrum sociocultural theories that propose sexually harassing behaviours are one of many consequences of gendered power inequalities (Fitzgerald et al., 1997; McDonald, 2012, 2016; Pina et al., 2009). Natural/biological theories will not be reviewed here because they have been dismissed for lack of depth and rigour to explain same-gender sexual harassment and for a general lack of empirical support (McDonald, 2016; Pina et al., 2009). Theories with greater rigour, depth and empirical support are organisational and feminist, each of which will be reviewed briefly.

Theories from an organisational perspective suggest that sexual harassment derives from organisational characteristics, specifically the job-gender context, workplace culture and differential worker power (McDonald, 2012). Job-gender context referred to the ratio of men and women and to the gendered characteristics of the job. Workplace culture described aspects of organisations (i.e. enforcing sanctions, taking complaints seriously and risk associated with making a complaint) that conveyed tolerance for sexually harassing behaviours. A meta-analysis of 41 studies examining 58 independent samples with a total sample size of nearly 70,000 provided extensive evidence that job-gender context and workplace culture facilitates sexual harassment (Willness et al., 2007). Chamberlain et al. (2008) coded 110 ethnographies conducted in a variety of industries (e.g. manufacturing, public administration, wholesale and retail trade) that included a range of occupations (e.g. farm, professional, assembly) to look at the effects of worker power on sexual harassment experiences. The authors defined worker power as self-direction (e.g. higher organisational status and power, skills and experiences that are difficult to replace), formal grievance procedures and job insecurity, with the findings showing statistically significant positive associations for self-direction, job insecurity and sexual harassment and a statistically significant negative associate for grievance procedures and sexual harassment. While organisational theories have provided clear, concise models to explain how organisational characteristics lead to sexual harassment, they tended to assume behaviours are readily addressed within the bubble of an organisation (Samuels, 2003), effectively ignoring broader sociocultural influences of gender regimes (Hearn & Parkin, 2003).

Feminist theories place gender and power relations at its core, with gender referring to a set of social relations imbued with a hierarchy of power (Walby, 2009). In Western cultures, it is generally accepted that there are two genders with which most people identify – man and woman – alongside an increasing recognition that there are multiple gender identities (e.g. Richards et al., 2016). The gender of men takes shape within gender relations and is understood to be a social category constructed in relation and in opposition to the gender of women (Connell, 2005). One means by which men construct their gender is the use of violence towards women (and other men), which allows them to maintain their power and dominance (Connell, 2005). "Violence is a part of the system of domination" (Connell, 2005, 84), as it allowed men to "draw boundaries and make exclusions" between them and women, as well as amongst men (Connell, 2005, 83). Sexual harassment is one of many manifestations of violence available in the process of power and dominance (re)construction. Supporting feminist theories are the decades of studies that show men are the primary perpetrators of sexual harassment, and women are the primary victim-survivors (Hunt et al., 2007; McDonald, 2012; Pina et al., 2009). Recently, feminist studies have been critiqued for their nearly exclusive focus on men sexually harassing women and for their heterosexist bias, rendering invisible men's experiences of victimisation and the influence of marginalised sexual orientations (McDonald, 2012). There is a small but growing body of studies documenting men's experiencing of sexual harassment from men and women colleagues and managers (e.g. Berdahl et al., 1996; McDonald, 2016), which indicates the foundation exists for sexual harassment studies to develop more nuanced understandings of workplace interactions and gendered power dynamics.

An area of overlap between sexual harassment and bullying theorising is that both have not engaged fully with the nuances of gendered power dynamics. Indeed, workplace bullying has been referred to as a field lacking firm theoretical foundation (Einarsen, 2000) and in need of an overarching, encompassing theory (Wheeler et al., 2010), yet these claims should not discount the conceptual insights that have developed from utilising other theories (Branch et al., 2013). Most bullying research is from the perspective of the individual victim-survivor. Recently, there has been expansion in bullying research to include group dynamics, organisational characteristics and societal influences (Branch et al., 2018). Group theories are an extension of individual theories as groups can be thought of as the immediate social environment from which individuals may learn bullying behaviours. Groups may consist of individuals with shared biographical characteristics, professional status or interests and individuals who do not conform to group norms or are marginalised in society by their social position. These individuals may be more likely to experience bullying from workplace groups, for example. Archer (1999). Organisational characteristics influencing bullying behaviours are organisational environments (i.e. changes in environments, such as restructuring and insecure employment, negative and stressful environments, poorly organised work environments); organisational culture (i.e. characterised by a high degree of conformity, group pressure); absence of clear organisational policies, leadership and management style (i.e. authoritarian style of settling disagreements, laissez-faire style of management); and power imbalances (Hoel & Salin, 2003; Samnani & Singh, 2012).

Bannerji (1995, 30, as cited in Berlingieri, 2015) extended theoretical understandings of workplace bullying beyond organisations:

> [The] workplace ... cannot be seen only as a place of economic production, but also must be understood as a coherent social and cultural environment which is organised through known and predictable social relations, practices, cultural norms and expectations.

Violence and gender are societal structures (Connell, 2005; Walby, 2009) that organise individual behaviour and workplace interactions (Hearn & Parkin, 2001) and influence manifestations of workplace bullying. Limiting workplace bullying as a distinct form of violence unconnected to gender, ethnicity and other social positions (e.g. sexual orientation, disability), the interrelated nature of workplace bullying with sexual harassment (and other forms of violence) remains hidden (Berlingieri, 2015). Moreover, not fully considering where and how workplace bullying and sexual harassment overlap runs the risk of curtailing victim-survivors' space to share their experiences, which may in turn disempower them and hinder intervention and prevention efforts.

Berlingieri (2015) drew from the continuum of violence model developed by Kelly (1987) to explore how workplace bullying and sexual harassment are interrelated. Key to the concept of the continuum of violence is

> that there are no clear boundaries separating forms of violence and the existence of common threads that weave through the various forms; that is, forms of violence possess both unique characteristics as well as characteristics shared with other forms of violence. They are interrelated, but not hierarchically.
> (Berlingieri, 2015, 347)

The shared characteristics are inequalities in power and control (Kelly, 1987), which is consistent with the theoretical understanding of sexual harassment and bullying (Branch & Murray, 2015; McDonald, 2012; McLaughlin et al., 2012). Even so, the question remains, is sexual harassment a form of bullying? Alternatively, it could be asked, is bullying a form of WGBV? The short answer to the questions is, it is complicated. These questions about the gendered nature of bullying and the relationships between bullying and sexual harassment need untangling, particularly as current understandings are based on traditional workplace power structures (Jones, 2006; Keashly, 2012) whereas cultural gigging economies rarely are.

Music festival WGBV

Music festivals are one of the original gig economies, whose organisational structures tend to reflect project-based organisations (PBO) as opposed to traditional organisational structures (Sedita, 2008). A PBO is a temporary social network of groups of people and organisations working together for a limited time to achieve a complex task. In this instance, the complex task is a live music event (Hartman et al., 1998; Sedita, 2008). The social network of employees needed to put on live music events at festivals can be grouped according to roles: festival manager, operations manager, box office and ticketing, finance and account management, promoters, technical and maintenance crew (e.g. sound engineer, stage builders), administration, event coordinators, stage managers and musicians (Rutter, 2016). Individuals performing these roles may not be found at every music festival, but when present, the specific employee performing the role tends to change from festival to festival and from stage to stage within a festival (Rutter, 2016). The hierarchy of power amongst employees is rarely defined and documented, so it would be difficult to classify certain job roles as management, as is done in traditional work organisations. The project management of festivals tends to reflect a strong focus on recreating musicians' intangible product of songs and non-musician employees greatly influencing the final product; the quality of the product is determined by the audience, whose perceptions are influenced by a host of factors (i.e. there are no objective standards for assessing quality) (Hartman, 1998). A music festival as an organisation is complex, often with multiple stages of musicians performing songs and different networks of employees working at each stage and employees dedicated to the overall running of the festival. The fluid and dynamic organisational structures of music festivals is probably the underlying reason why research on sexual harassment and bullying in music festivals could not be located. It could be difficult to identify and recruit music festival employees to take part in research. Nonetheless, there is evidence supporting the presence of proxy indicators.

A brief review of research evidencing organisational and job proxy indicators is provided here, with the acknowledgement that other factors may cause these negative workplace behaviours in festivals. Organisational proxies are characteristics of organisations that have been found to be associated with sexual harassment and bullying. Both fields have demonstrated that the job-gender context and the climate of organisations are substantial and significant proxies. The job-gender context refers to the ratio of men to women in an organisation. When the ratio favours men, sexual harassment and bullying are more likely to occur (e.g. Archer, 1999; Chamberlain et al., 2008; Einarsen, 2000; Gruber, 1998). A recent study of nine festivals' posters advertising bands found that the gender ratio favoured men, with festival line ups consisting of 77% all-male bands, 13% all-female bands and 10% male-and-female bands (BBC Reality Check Team, 2018). This indicates music festivals primarily advertise and sell the creative product of male musicians. The male-dominated gender ratio continues across other job roles in live music events (Webster et al., 2018). Taken together, outward-facing employees directly providing a creative product – i.e. musicians – and the inward-facing employees shaping the product are predominantly men, which means there is a gendered power imbalance amongst music festival employees. The power differential is a key facilitator for male employees to use sexually harassing and bullying behaviours at music festivals. The organisational climate has several components, including the poorly organised environments, absence of clear leadership styles and lack of formal policies and grievance procedures. As said earlier, music festivals are characterised by temporary networks of employees with hierarchies of power varying across festivals and stages within festivals. The potential for inconsistent organisation within and across music festivals, as well as the potential for variation in leadership styles, creates opportunities for negative workplace behaviours. Policies and procedures clearly setting out acceptable workplace behaviour and consequences for unacceptable behaviour serve as a deterrent for perpetrators and provide recourse for victim-survivors (Chamberlain et al., 2008; Willness et al., 2007). Approximately two-thirds of live music event organisations lacked complaint procedures and policies (Webster et al., 2018), suggesting that there are increased chances for employees to perpetrate and fewer opportunities for the needs of victim-survivors to be addressed in the workplace. Job proxy indicators describe characteristics of the job role itself, such as job security. Employees with insecure employment are more likely to experience negative workplace behaviours (e.g. Baillien & De Witte, 2009; Notelaers et al., 2010). Most employees are self-employed, relying on networks for future work (Rutter, 2016; Sedita, 2008); they may work for very little or no pay (Gross & Musgrave, 2016); thus, they may be at particularly high risk for sexual harassment and bullying.

Manifestations of WGBV and impacts

In line with proxy indicators suggesting that sexual harassment and bullying occur amongst music festival employees, anecdotal evidence (e.g. Mackenzie, 2017) and consultations with stakeholders suggested sexual harassment and bullying are widespread. This informal evidence reflected the popular understanding that men with more organisational power and authority sexually harass women and bully men and women. This understanding amongst employees needs to be broadened, as this is only one of the perpetrator victim-survivor relationship types identified in the literature. Others include both genders using negative workplace behaviours towards women and men of the same professional standing (Jones, 2006), men perpetrating negative workplace behaviours towards women with more power (Jones, 2006), multiple employees bullying one person (e.g. Török et al., 2016) and audience members – i.e. music festival customers – harassing or bullying employees (e.g. Equalities and Human Rights Commission, 2018; Friborg et al., 2017; Yagil, 2018). The specific behaviours that are used amongst music festival employees and music festival customers could be any of the behaviours that fall into the general understanding of sexual harassment and bullying. The range from an inappropriate comment about an individual's sexual activities to repeated comments belittling and undermining an individual's job performance to physical violence to inappropriate touching of a sexual nature.

It is important to keep in mind the associated consequences experienced by victim-survivors and employees who witness sexually harassing and bullying behaviours, as they may manifest in festival workplaces. Victim-survivors and witnesses may experience post-traumatic stress disorder, anxiety and depression and decreased physical well-being (e.g. Willness et al., 2007; Matthiesen & Einarsen, 2010; McDonald, 2012). These consequences are associated with decreased job performance. They may experience other job-related consequences, such as decreased job satisfaction, absenteeism and looking for work elsewhere (e.g. Willness et al., 2007; Matthiesen & Einarsen, 2010). Additionally, victimisation outside of work can and does influence employment settings. GBV victimisation can lead directly or indirectly to a multitude of health consequences similar to those described earlier for GBV experienced in the workplace (Jina & Thomas, 2013). Consequences may result immediately or in the medium to long term. Examples of immediate consequences include acute physical injuries, sexually transmitted diseases, pregnancy and genital or anal trauma; and examples of medium- to long-term consequences include gastrointestinal symptoms, chronic pain, infertility, sleeping difficulty and increased use of drugs and alcohol (Campbell et al., 2009; WHO, 2013). Additional areas that may be affected include (but are not limited to) economic and social well-being. Loya (2015) interviewed 27 victim-survivors and service providers about the economic consequences, with the majority describing victim-survivors taking time off work, performing less well in work and changing career paths. A recent report by the UK Home Office (Oliver et al., 2019) estimated that victim-survivors of intimate partner rape lost 138 hours (approximately 1 month) of paid work, and when in work, their productivity was reduced for 552 hours (approximately 3.7 months). The time off work and reduced productivity equated to a loss of £75,990 in 2016 and 2017. Disclosures of victimisation within and outside workplaces are more likely to be made to informal sources of support, such as friends, family and co-workers (e.g. Sinha, 2013; ONS, 2018), and these disclosures can lead to informal social support providers experiencing physical and mental health consequences (e.g. Ahrens & Campbell, 2000; Gregory et al., 2017; Gregory et al., 2017; Riger et al., 2002). It is reasonable to assume that the employment of friends, family and co-workers who are impacted by disclosures might also experience consequences in their work and well-being.

Research and policy recommendations

Workplace violence has been recognised internationally by the European Union, the United Nations Committee on the Elimination of Discrimination Against Women and the World Health Organisation, and recently, the International Labour Organisation in 2019 adopted a global treaty. In response to the growing international recognition, organisations created policies and reporting systems and trained employees (McCann, 2005). In the UK in 2017, more than 60 members of the Association for Independent Festivals signed a charter in which they agreed to a zero-tolerance stance towards sexual harassment (Chapple, 2017). While an important step to initiate changes in festival cultures, the underlying focus was festival attendees experiencing sexual harassment, not employees. Policies and reporting procedures tailored to the unique work environment of music festivals need to be developed that clearly define not only sexual harassment but also bullying in a manner that does not silo each from the other. Policies should also set out potential work-related consequences and legal ramifications for employees. Raising awareness amongst employees of the policies and procedures specific to them may prove difficult as most are self-employed individuals working without contractual obligations for festivals or bands. In other words, it may be challenging to identify all employees working at a music festival site. Nonetheless, every effort should be made to reach as many employees as possible to establish the cultural norm negative workplace behaviours will not be tolerated, and if and when they occur, employees will be sanctioned.

Bystander education and prevention programmes are an effective tool to raise awareness and to teach techniques to support victim-survivors, respond to perpetrators in a safe manner and shift cultural norms supporting violent and abusive behaviours (Jouriles et al., 2018). Additionally, they have been shown to reduce perpetration (e.g. Coker et al., 2015). Bystander programmes specific to music festival workplaces need to be developed. Development needs to include research on sexual harassment and bullying in music festival workplaces because programmes with content directly related to the social context in which it is implemented have been shown to be more effective (Nation et al., 2003). Once developed, programmes should be implemented as extensively as possible. Initial implementation could be on a smaller scale by engaging male employees well-known throughout music festivals and high-profile male bands. Male allies are critical to challenge cultural norms that support violent and abusive behaviours (Flood, 2004).

The gender ratio of music festival employees needs to be addressed. Acknowledgement of gender inequalities is limited (Bennett, 2018) but growing with media articles on gender disparities in musicians booked to play at live events (e.g. Cooper, 2019). The live events industry has responded with schemes to reduce the disparity amongst employees (e.g. Cooke, 2018). The schemes introduce various job roles to women and provide training. These efforts indicate that there is a readiness to change in the industry. However, schemes aimed at bringing women into the music festival workplace need to be supplemented with research on if and why women leave their jobs. Underlying this recommendation is research that showed employees who experienced sexual harassment or bullying were more likely to look for jobs elsewhere and or change work industries. It is essential for music festivals to understand how their power structures and characteristics facilitate sexual harassment and bullying in order for cultural changes to be made which will support women to continue to work in music festivals.

Music festivals taking up these recommendations would demonstrate that they are taking steps to be in line with the Equalities and Human Rights Commission (2018) and the Women and Equalities Committee (2018), both of which called for better prevention of and responses to violent workplace behaviours. UK music festivals are at a pivotal crossroad to answer these calls in which they can fully take on the recommendations laid out here to create environments that are free of the constraints found elsewhere in life ... or they can maintain the status quo.

References

Ahrens, C. E., & Campbell, R. (2000). Assisting rape victims as they recover from rape: The impact on friends. *Journal of Interpersonal Violence, 15*, 959–986.

Archer, D. (1999). Exploring "bullying" culture in the para-military organisation. *International Journal of Manpower, 20*, 94–105.

Baillien, E., & de Witte, H. (2009). Why is organizational change related to workplace bullying? Role conflict and job insecurity as mediators. *Economic and Industrial Democracy, 30*, 348–371.

Bannerji, H. (1995). *Thinking through: Essays on Feminism, Marxism and Anti-Racism*. Toronto, ON, Canada: Canadian Scholars' Press and Women's Press.

Bennett, T. (2018). The whole feminist taking-your-clothes-off thing: Negotiating the critique of gender inequality in UK music industries. *Journal of the International Association for the Study of Popular Music, 8*(1), 24–41.

BBC Reality Check Team. (2018). Festivals 2018: The gender gap in music festival line-ups. *BBC* [Online]. Retrieved September 30, 2019, from www.bbc.co.uk/news/entertainment-arts-44655719

Berdahl, J. L., Magley, V. J., & Waldo, C. R. (1996). The sexual harassment of men? Exploring the concept with theory and data. *Psychology of Women Quarterly, 20*, 527–547.

Berlingieri, A. (2015). Workplace bullying: Exploring an emerging framework. *Work, Employment and Society, 29*, 342–353.

Black, M., Basile, K., Breiding, M., Smith, S., Walters, M., Merrick, M., Chen, J., & Stevens, M. (2011). *National Intimate Partner and Sexual Violence Survey (NISVS): 2010 summary report*. Altanta, George: National Ctr for Injury Prevention and Control, and Center for Disease Control, Division of Violence Prevention.

Branch, S., Linda, S., Barker, M., Ramsay, S., & Murray, J. P. (2018). Theoretical frameworks that have explained workplace bullying: Retracing contributions across the decades. In P. D'Cruz, E. Noronha, G. Notelaers, & C. Rayner (Eds.), *Handbook of workplace bullying, emotional abuse, and harassment, Volume 1: Concepts, approaches, and methods*. Singapore: Springer Nature.

Branch, S., & Murray, J. (2015). Workplace bullying. *Organizational Dynamics, 44*, 287–295.

Branch, S., Ramsay, S., & Barker, M. (2013). Workplace bullying, mobbing and general harassment: A review. *International Journal of Management Reviews, 15*, 280–299.

Campbell, R., Dworkin, E., & Cabral, G. (2009). An ecological model of the impact of sexual assault on women's mental health. *Trauma, Violence, & Abuse, 10*, 225–246.

Chamberlain, L. J., Crowley, M., Tope, D., & Hodson, R. (2008). Sexual harassment in organizational context. *Work and Occupations, 35*, 262–295.

Chapple, J. (2017). UK festivals sign anti-sexual harassment charter. *IQ: Live Music Intelligence* [Online]. Retrieved January 28, 2018, from www.iq-mag.net/2017/05/uk-festivals-sign-safer-spaces-charter/#.XXJzhTZKhyz

Chapple, J. (2018). Live music revenues to top £30bn for first time. *IQ: Live Music Intelligence* [Online]. Retrieved March 17, 2019, from www.iq-mag.net/2018/10/live-music-revenues-to-top-30bn-for-first-time/#.XXI7tjZKhyw

Coker, A. L., Fisher, B. S., Bush, H. M., Swan, S. C., Williams, C. M., Clear, E. R., & Degue, S. (2015). Evaluation of the green dot bystander intervention to reduce interpersonal violence among college students across three campuses. *Violence against Women*, *21*(12), 1507–1527.

Connell, R. W. (2005). *Masculinities* (2nd ed.). Cambridge, UK: Polity Press.

Cooke, C. (2018). Music venue trust launches scheme to support new female gig promoters. *Complete Music Update* [Online]. Retrieved July 1, 2019, from https://completemusicupdate.com/article/music-venue-trust-launches-scheme-to-support-new-female-gig-promoters/

Cooper, L. (2019). Girls to the front: Why gender is still a headline issue at festivals. *The Guardian* [Online]. Retrieved June 4, 2019, from www.theguardian.com/music/2019/may/03/girls-to-the-front-why-gender-is-still-a-headline-issue-at-festivals

Coulter, R. W., & Rankin, S. R. (2017). College sexual assault and campus climate for sexual- and gender-minority undergraduate students. *Journal of Interpersonal Violence*. doi:10.1177/0886260517696870

Einarsen, S. (2000). Harassment and bullying at work: A review of the Scandinavian approach. *Aggression and Violent Behavior*, *5*, 379–401.

Equality and Human Rights Commission. (2018). Turn the table: Ending sexual harassment at work. *Equality and Human Rights Commission* [Online]. Retrieved July 17, 2019, from www.equalityhumanrights.com/sites/default/files/ending-sexual-harassment-at-work.pdf

Farley, L. (1978). *Sexual shakedown: The sexual harassment of women in the working world*. London: Melbourne House.

Fileborn, B., Wadds, P., & Tomsen, S. (2018). New research shines light on sexual violence at Australian music festivals. *The Conversation* [Online]. Retrieved September 30, 2019, from https://theconversation.com/new-research-shines-light-on-sexual-violence-at-australian-music-festivals-104768

Fitzgerald, L. F., Drasgow, F., Hulin, C. L., Gelfand, M. J., & Magley, V. J. (1997). Antecedents and consequences of sexual harassment in organizations: A test of an integrated model. *Journal of Applied Psychology*, *82*, 578–589.

Flood, M. (2004). *Changing men: Best practice in violence prevention work with men*. Home Truths Conference: Stop Sexual Assault and Domestic Violence: A National Challenge, Melbourne, September 15–17.

Friborg, M. K., Hansen, J. V., Aldrich, P. T., Folker, A. P., Kjær, S., Nielsen, M. B. D., Rugulies, R., & Madsen, I. E. (2017). Workplace sexual harassment and depressive symptoms: A cross-sectional multilevel analysis comparing harassment from clients or customers to harassment from other employees amongst 7603 Danish employees from 1041 organizations. *BMC Public Health* [Online], *17*, 675. doi:10.1186/s12889-017-4669-x

Garcia-Moreno, C., Jansen, H. A., Ellsberg, M., Heise, L., & Watts, C. H. (2006). Prevalence of intimate partner violence: Findings from the WHO multi-country study on women's health and domestic violence. *The Lancet*, *368*, 1260–1269.

Gregory, A. C., Feder, G., Taket, A., & Williamson, E. (2017). Qualitative study to explore the health and well-being impacts on adults providing informal support to female domestic violence survivors. *BMJ Open* [Online], *7*, e014511. doi:10.1136/bmjopen-2016-014511

Gregory, A. C., Williamson, E., & Feder, G. (2017). The impact on informal supporters of domestic violence survivors: A systematic literature review. *Trauma, Violence, & Abuse*, *18*, 562–580.

Griffin, C., Bengry-Howell, A., Riley, S., Morey, Y., & Szmigin, I. (2018). "We achieve the impossible": Discourses of freedom and escape at music festivals and free parties. *Journal of Consumer Culture*, *18*, 477–496.

Gross, S., & Musgrave, G. (2016). Can music make you sick part 1? A study into the incidence of musicians' mental health. *Middlesex: Music Tank Publishing* [Online]. Retrieved June 7, 2018, from www.helpmusicians.org.uk/assets/publications/files/can_music_make_you_sick_part_1-_pilot_survey_report_2019.pdf

Gruber, J. E. (1998). The impact of male work environments and organizational policies on women's experiences of sexual harassment. *Gender & Society*, *12*, 301–320.

Hartman, F., Ashrafi, R., & Jergeas, G. (1998). Project management in the live entertainment industry: What is different? *International Journal of Project Management*, *16*, 269–281.

Hearn, J., & Parkin, W. (2001). *Gender, sexuality and violence in organizations: The unspoken forces of organization violations*. London: Sage Publications.

Hearn, J., & Parkin, W. (2003). The gendered organisation: A positive critique. *Comportamento organizacional e gestão*, *9*, 125–146.

Hester, M., Jones, C., Williamson, E., Fahmy, E., & Feder, G. (2017). Is it coercive controlling violence? A cross-sectional domestic violence and abuse survey of men attending general practice in England. *Psychology of Violence*, *7*, 417–427.

Hester, M., & Lilley, S.-J. (2014). *Preventing violence against women: Article 12 of the Istanbul convention: A collection of papers on the council of Europe convention on preventing and combating violence against women and domestic violence* [Online]. Strasbourg, France: Council of Europe. Retrieved from September 4, 2016, from https://research-information.bristol.ac.uk/files/34955886/Article_12_English.pdf

Hoel, H., & Salin, D. (2003). Organisational antecedents of workplace bullying. In S. Einarsen, H. Hoel, D. Zapf, & C. L. Cooper (Eds.), *Bullying and harassment in the workplace: Developments in theory, research, and practice* (pp. 203–218). Baco Raton, FL: CRC Press.

Hunt, C., Davidson, M., Fielden, S., & Hoel, H. (2007). *Sexual harassment in the workplace: A literature review* [Online]. Manchester: Equal Opportunities Commission, Working Paper Series. Retrieved May 6, 2019, from https://research.mbs.ac.uk/equality-diversity/Portals/0/docs/WPS59Sexualharassment.pdf

Jagsi, R., Griffith, K. A., Jones, R., Perumalswami, C. R., Ubel, P., & Stewart, A. (2016). Sexual harassment and discrimination experiences of academic medical faculty. *Journal of the American Medical Association*, *315*, 2120–2121.

Jina, R., & Thomas, L. S. (2013). Health consequences of sexual violence against women. *Best Practice & Research Clinical Obstetrics and Gynaecology*, *27*, 15–26.

Jones, C. (2006). Drawing boundaries: Exploring the relationship between sexual harassment, gender and bullying. *Women's Studies International Forum*, *29*, 147–158.

Jones, C. (2019). Gender, power, and powerlessness: A conceptual framework for researching men's victimisation. In H. Pontell (Ed.), *Oxford research encyclopedia of criminology and criminal justice*. New York and Oxford: Oxford University Press. doi: http://dx.doi.org/10.1093/acrefore/9780190264079.013.535

Jouriles, E. N., Kraus, A., Vu, N. L., Banyard, V. L, & Mcdonald, R. (2018). Bystander programs addressing sexual violence on college campuses: A systematic review and meta-analysis of program outcomes and delivery methods. *Journal of American College Health*, *66*(6), 457–466.

Keashly, L. (2012). Workplace bullying and gender: It's complicated. In S. Fox & T. R. Lituchy (Eds.), *Gender and the dysfunctional workplace* (pp. 78–95). Cheltenham, UK: Edward Elgar Publishing Limited.

Kelly, L. (1987). The continuum of sexual violence. In M. Maynard & J. Hanmer (Eds.), *Women, violence and social control* (pp. 46–60). London: Palgrave Macmillan.

Klatt, H.-J. (2018). Sexual harassment policies as all-purpose tools to settle conflicts. In B. M. Dank & R. Refinetti (Eds.), *Sexual harassment and sexual consent* (pp. 45–69). Abingdon, Oxford and New York, NY: Routledge.

Long, M. (2018). Unwanted sexual behaviours "rife" at UK festivals. *IQ: Live Music Intelligence* [Online]. Retrieved October 3, 2019, from www.iq-mag.net/2018/06/sexual-assault-harassment-rife-uk-festivals/#.XZnalvlKhyz

Loya, R. M. (2015). Rape as an economic crime: The impact of sexual violence on survivors' employment and economic well-being. *Journal of Interpersonal Violence*, *30*, 2793–2813.

Mackenzie, J. (2017). Rape and abuse: The music industry's dark side exposed. *BBC* [Online]. Retrieved February 3, 2018, from www.bbc.co.uk/news/entertainment-arts-42368544

Mackinnon, C. A. (1979). *Sexual harassment of working women: A case of sex discrimination*. New Haven and London: Yale University Press.

Matthiesen, S. B., & Einarsen, S. (2010). Bullying in the workplace: Definition, prevalence, antecedents and consequences. *International Journal of Organization Theory & Behavior*, *13*, 202–248.

Mccann, D. (2005). *Sexual harassment at work: National and international responses. Project Report*. Geneva: International Labour Organization.

Mcdonald, P. (2012). Workplace sexual harassment 30 years on: A review of the literature. *International Journal of Management Reviews*, *14*, 1–17.

Mcdonald. P. (2016). Workplace sexual harassment at the margins. *Work, Employment & Society*, *30*(1), 118–134.

Mclaughlin, H., Uggen, C., & Blackstone, A. (2012). Sexual harassment, workplace authority, and the paradox of power. *American Sociological Review*, *77*, 625–647.

Morgan, M. (2007). Festival spaces and the visitor experience. In M. Casado-Diaz, S. Everett, & J. Wilson (Eds.), *Social and cultural change: Making spaces(s) for leisure and tourism* (pp. 113–130). Eastbourne, UK: Leisure Studies Association.

Nation, M., Crusto, C., Wandersman, A., Kumpfer, K. L., Seybolt, D., Morrissey-Kane, E., & Davino, K. (2003). What works in prevention: Principles of effective prevention programs. *American Psychologist*, *58*, 449–456.

Notelaers, G., de Witte, H., & Einarsen, S. (2010). A job characteristics approach to explain workplace bullying. *European Journal of Work and Organizational Psychology*, *19*, 487–504.

Office of National Statistics (ONS). (2018). *Sexual offences in England and Wales year ending March 2017* [Online]. Retrieved July 22, 2019, from www.ons.gov.uk/peoplepopulationandcommunity/crimeandjustice/articles/sexualoffencesinenglandandwales/yearendingmarch2017#how-are-sexual-offences-defined-and-measured

Oliver, R., Alexander, B., Roe, S., & Wlasny, M. (2019). *The economic and social costs of domestic abuse* [Online]. London: Home Office. Retrieved July 30, 2019, from https://assets.publishing.service.gov.uk/government/uploads/system/uploads/attachment_data/file/772180/horr107.pdf

Pina, A., Gannon, T. A., & Saunders, B. (2009). An overview of the literature on sexual harassment: Perpetrator, theory, and treatment issues. *Aggression and Violent Behavior, 14*, 126–138.

Richards, C., Bouman, W. P., Seal, L., John Baker, M. Nielder, T. O., & T'Sjoen, G. (2016). Non-binary or genderqueer genders. *International Review of Psychiatry, 28*, 95–102. doi:10.3109/09540261.2015.1106446

Riger, S., Raja, S., & Camacho, J. (2002). The radiating impact of intimate partner violence. *Journal of Interpersonal Violence, 17*, 184–120.

Rutter, P. (2016). *The music industry handbook*. Abingdon, Oxford: Routledge.

Samnani, A.-K., & Singh, P. (2012). 20 years of workplace bullying research: A review of the antecedents and consequences of bullying in the workplace. *Aggression and Violent Behavior, 17*, 581–589.

Samuels, H. (2003). Sexual harassment in the workplace: A feminist analysis of recent developments in the UK. *Women's Studies International Forum, 26*, 467–482.

Sedita, S. R. (2008). Interpersonal and inter-organizational networks in the performing arts: The case of project-based organizations in the live music industry. *Industry and Innovation, 15*, 493–511.

Sinha, M. (2013). *Measuring violence against women: Statistical trends* [Online]. Ottawa, Ontario: Canadian Centre for Justice Statistics. Retrieved June 16, 2015, from www150.statcan.gc.ca/n1/pub/85-002-x/2013001/article/11766-eng.pdf

Stöckl, H., Devries, K., Rotstein, A., Abrahams, N., Campbell, J., Watts, C., & Moreno, C. G. (2013). The global prevalence of intimate partner homicide: A systematic review. *The Lancet, 382*, 859–865.

Szmigin, I., Bengry-Howell, A., Morey, Y., Griffin, C., & Riley, S. (2017). Socio-spatial authenticity at co-created music festivals. *Annals of Tourism Research, 63*, 1–11.

Tinkler, J., Gremillion, S., & Arthurs, K. (2015). Perceptions of legitimacy: The sex of the legal messenger and reactions to sexual harassment training. *Law & Social Inquiry, 40*, 152–174.

Török, E., Hansen, Å. M., Grynderup, M. B., Garde, A. H., Høgh, A., & Nabe-Nielsen, K. (2016). The association between workplace bullying and depressive symptoms: The role of the perpetrator. *BMC Public Health, 16*, 993–1030.

UK Music. (2018). *Measuring Music* [Online]. London: UK Music. Retrieved February 22, 2019, from www.ukmusic.org/assets/general/UK_Music_Measuring_Music_2018.pdf

Vartia, M., & Hyyti, J. (2002). Gender differences in workplace bullying among prison officers. *European Journal of Work and Organizational Psychology, 11*, 113–126.

Walby, S. (2009). *Globalization and inequalities: Complexity and contested modernities*. London: Sage Publications.

Watts, C., & Zimmerman, C. (2002). Violence against women: Global scope and magnitude. *The Lancet, 359*, 1232–1237.

Webster, E., Brennan, M., Behr, A., Cloonan, M., & Ansel, J. (2018). *Valuing live music: The UK live music census 2017 report*. Edinburgh: University of Edinburgh and Live Music Exchange.

Wheeler, A. R., Halbesleben, J. R., & Shanine, K. (2010). Eating their cake and everyone else's cake, too: Resources as the main ingredient to workplace bullying. *Business Horizons, 53*, 553–560.

Wilks, L. (2011). Bridging and bonding: Social capital at music festivals. *Journal of Policy Research in Tourism, Leisure and Events, 3*, 281–297.

Willness, C. R., Steel, P., & Lee, K. (2007). A meta-analysis of the antecedents and consequences of workplace sexual harassment. *Personnel Psychology, 60*, 127–162.

Women and Equalities Committee. (2018). *Sexual harassment in the workplace* [Online]. London: House of Commons. Retrieved January 27, 2019, from https://publications.parliament.uk/pa/cm201719/cmselect/cmwomeq/725/725.pdf

World Health Organization. (2013). *Responding to intimate partner violence and sexual violence against women: WHO clinical and policy guidelines* [Online]. World Health Organization. Retrieved November 18, 2015, from www.who.int/iris/bitstream/10665/85240/1/9789241548595_eng.pdf?ua=1

Yagil, D. (2018). Abuse from organizational outsiders: Customer aggression and incivility. In P. D'Cruz, E. Noonha, L. Keashly, & S. Tye-Williams (Eds.), *Special topics and particular occupations, professions and sectors* (pp. 1–26). Singapore: Springer Nature.

9 Structural disputes

An analysis of infrastructural inequalities in the case study of the Women of the World Festival in Hull (UK) City of Culture 2017

Barbara Grabher

Introduction

In 2017, Hull, a medium-sized city in the north-east of England, celebrated being named the UK City of Culture (UKCOC) through "365 days of transformative culture" (Hull 2017 Ltd, 2015, 14). The celebrations of arts, culture and heritage aspired economic, social and cultural regeneration, as the city of Hull aims to move out of the shadows of deprivation. With over 2,800 events, the schedule was full of opportunities to engage with and explore the multiple facets of creativity that the city, region and country had to offer (Culture Place and Policy Institute, 2018). While not formally outlined as a central ambition of the project, the year-long cultural programme foregrounded on multiple occasions the struggle for gender equality and social justice as a core value of Hull's UKCOC award. Large-scale festivals, such as the commemorative celebrations of LGBT50 and the annually recurring Freedom Festival, fostered a culture of equality that the celebratory community shared. Another crucial celebration of gender equality took place from March 10–12, 2017, as the Women of the World (WOW) festival claimed Hull City Hall. WOW Hull promised to "celebrate women and girls and take a frank look at what stops them from achieving their potential" (Hull 2017 Ltd, 2017d).

In an ethnographic study of the event, I collected insights from producers, artists and visitors at the festival in order to understand how the WOW festival negotiates the notion of equality on an experiential, discursive and infrastructural level. While the experience and content of the festival foreground ambitious engagement with equality, in this chapter, I highlight structural disputes over equality as an infrastructural condition. I outline the limitations and difficulties of the infrastructural conditions of the WOW festival in Hull in 2017. Through an analysis of structural disputes on the micro-, meso- and macro-levels, I argue for the fragility of equality. Inadequate structures put the discursive negotiations of the concept at risk. I do not aim to scrutinise WOW Hull, nor any projects affiliated to the festival franchise. Rather, I want to highlight the need to celebrate equality within adequate structures. The celebration of equality as a political struggle is crucial and needs to be considered as a positive development.

120 *Barbara Grabher*

However, ongoing commodification risks compromising the achievements of the equality movement.

The chapter is divided into three sections. First, I situate the investigation in reference to the relevant research affiliations, field and methodology. Second, I engage with the conceptual discussions associated with the celebrations of equality. While acknowledging the potential of celebrations as sociocultural practices of meaning-making, I engage with questions concerning structural violence and its particular application to festival contexts. Third, I illustrate micro-, meso- and macro-encounters of structural disputes, as noted by research participants at the WOW festival in Hull. The perceived restrictions imposed by the associated labels, the establishment of restricted access due to ticketing systems and VIP spaces and the tendencies of commodification through corporate festival branding are three examples that highlight inherent structural inequalities within the celebration of equality in the WOW festival in Hull in 2017.

Situating the research

Through brief discussions of the relevant institutional affiliations, the circumstances of the field and the methodological approach, I situate the research in its scholarly context.

Institutional affiliation

The material presented in this chapter derives from the research project *Gendering Cities of Culture*. In this project, I investigate the production of cultures of equality in the mega-events of Hull UKCOC 2017 (Hull2017) and Donostia/San Sebastián European Capital of Culture (ECOC) 2016 (DSS2016). In the overarching project, I explore the production of cultures of equality in the two field sites through six selected activities in the "365 days of transformative culture" (Hull 2017 Ltd, 2015, 14). Due to limitations of space, this chapter concentrates on the case study of the WOW festival in Hull in association with the city's celebration of the UKCOC award. *Gendering Cities of Culture* forms part of the *GRACE Project* (Gender and Cultures of Equality in Europe) and is funded through the European Commission *Horizon 2020 Marie Skłodowska-Curie Actions* framework.

Field

In this research, the UKCOC initiative, the City of Hull, the mega-event of Hull2017, and the particularities of the WOW festival in Hull interlink and constitute my research field. Due to these prominent entanglements, the research requires me to address the field in accordance with Hilgers and Mangez's (2015) interpretation of Bourdieu's notion of field. Rather than a geo-spatial unit, the field is characterised as an entity of relations between geographies, institutions, and practices. By introducing the city of Hull, the UKCOC mega-event, and the WOW festival, I disentangle the most essential elements of the field for this study.

Kingston upon Hull, referred to as Hull, is situated in the county of Yorkshire and the Humber in the north-east of England, at the junction of the Humber Estuary and Hull River. The city counts around 260,000 inhabitants (Office for National Statistics, 2019). As a port city, its geography is oriented along the river banks and spreads into the hinterlands. The city was first mentioned in the twelfth century and named Kingston upon Hull under King Edward I in 1299. It gained importance in trade due to the access to the North Sea provided by the Humber Estuary. Trade connected the city with Scandinavia, the Baltic region and the Low Countries. At its height in the eighteenth and nineteenth centuries, the whaling and fishing industries influenced the urban expansion and progressions. The accumulated wealth and importance of the port city was the reason for major bombing during World War II in which 95% of the city centre was destroyed or damaged. In the 1950s and 1960s, the city recovered from the war in economic and social terms. While Hull experienced another decade of importance in trading and fishing, the 1970s saw a collapse for the city's industry. Ever since, the population has dealt with the socio-economic consequences (Hull City Council, n.d.). The preliminary evaluation report of Hull2017 summarises that, with an unemployment rate of 7% in 2016, Hull is the third-most deprived local authority out of the 326 local areas in England. The average gross value added (GVA) per head is as low as 18,000 GBP in Hull in comparison to the national GVA of 27,000 GBP per head (Culture Place and Policy Institute, 2018).

The bidding, selection and execution of the UKCOC is an essential element in the city's regeneration plan to overcome the socio-economic consequences of de-industrialisation. "A city coming out of its shadows" (Culture Place and Policy Institute, 2018, 6) served as motivation for the selection panel's decision in 2013 to grant the "badge of authority" (Redmond, 2009, 2) to Hull. While not affiliated with explicit monetary value, the title's attractiveness comes from the associated prestige of the award. Carried by the Department for Culture, Media and Sports (2013), the national title was inspired by Glasgow's and Liverpool's success as hosts of the ECOC in 1990 and 2008. The UKCOC award was first celebrated by Derry/Londonderry in 2013. Hull is the second city to be awarded the title UKCOC and is followed by Coventry in 2021.

Central to Hull's celebration was the programming of 365 days of cultural activities in 2017. Structured along four programming seasons entitled *Made in Hull*, *Roots and Routes*, *Freedom* and *Tell the World*, the programme included four major events, which engaged audiences with particular politically relevant topics. The WOW festival constituted the first of these focussed events (Hull 2017 Ltd, 2017a, 2017b, 2017c).

The multi-day, multi-platform festival took place in the City Hall and surrounding venues from March 10–12, 2017, in reference to International Women's Day. Affiliated with the WOW festival brand, Hull's celebration embeds in a global network of festivals for gender equality. The WOW festival franchise originates from Southbank Centre in London, where the festival brand was created by former director Jude Kelly. The festival's ambitions are explained as follows:

> Women of the World is a global movement celebrating women and girls, taking a frank look at the obstacles they face. [...] It is the biggest, most comprehensive and most significant festival dedicated to presenting work by women and promoting equality for women and girls.
>
> (Southbank Centre, 2018)

Since its inauguration in 2010, the festival has been celebrated nationally and internationally, as various WOW festivals take place on a semi-regular schedule in the UK and other countries, such as Brazil, Australia and the US. WOW Hull was organised by a team from Hull2017 Ltd and led by executive producer Henrietta Duckworth and programmer Madeleine O'Reiley. During the organisational process, the WOW Circle of Friends was established, which brought together local representatives of gender-specific institutions in and around the city. The programming of the festival was informed by a series of community brainstorming events entitled WOW Think-ins, which took place between September and November 2016 in different community hubs around the city. Inspired by conversations held with participants of the WOW Think-ins, the programme included an eclectic combination of talks, debates and performances by local representatives. During the festival, a central act was the awarding of trailblazer status to local women of the past and present who have contributed to a more just and equal society. The list includes artists and activists, such as Mary Wollstonecraft, the Headscarf Revolutionaries and Maureen Lipman.[1] The festival achieved a fairly low attendance in comparison to its original expectations. Even though informally discussed as a legacy project, WOW Hull was a one-off event and did not return in 2018.

Methodology

In the ethnographic study of the WOW festival, my research practice is driven by a feminist methodology of reflexivity. Rather than aspiring to objectivity, I embrace a commitment to social values as the methodological approach of this study. My considerations derive from Harding's (1993) call for strong objectivity. The scholar outlined, "Maximising objectivity in social research requires not total value neutrality, but instead, a commitment by the research[er] to certain values" (Hirsch & Olson, 1995, 202). In accordance with scholars of the emerging field of critical event studies (CES), I see this methodological approach applied in the study of festivals and events. Lamond and Platt (2016, 2) observed, "Present methodological discussion within event studies is often dominated by the changing demands for refinement of methods suitable for event evaluation". Countering objectivity-driven, neoliberal research agendas, their publication, *Critical Event Studies: Approaches to Research*, introduces scholars such as Pavoni and Citroni (2016), Dashper (2016) and Finkel and Sang (2016) in their discussion of innovative approaches in event research, as they encompass ethnographic, auto-ethnographic and participatory practices of investigations. This emerging and crucial research perspective supports me as an orientation in my methodological approach to the study of festivals.

The analysis of the infrastructural conditions of the WOW festival in Hull is based on my participatory observations in the WOW Circle of Friends, selected WOW Think-ins events and the three days of the festival. Additionally, I was able to conduct semi-structured interviews with one actor involved in the management of the event as well as three artists contributing to the programme of the three-day festival. Furthermore, I collected perceptive accounts of the festival from audience members attending the event. Hereby, I established a collaboration with a group of five observing- participants, who, as residents of Hull, visited, explored and observed multiple events as part of Hull2017. The team observing the WOW festival include three cis-women, one trans-woman and one man who cover an age range between 30 and 75 years old. Their location of residence spreads across the east, west and centre of the city (Grabher, 2018). The data were analysed through Mayring's (1994) suggested technique of qualitative content analysis. In the further presentation of the analysis, all references to participants are anonymised through pseudonyms.

Celebrating equality – conceptual considerations

The ambitions of the WOW festival are outlined as a celebration of equality. In order to investigate these aspirations, Finkel et al. (2013), as well as Finkel (2009, 2015), suggested considering festivals as sociocultural practices of meaning-making. In the edited volume *Time Out of Time*, Falassi (1987, 2) explained practices of meaning-making in celebrations as follows:

> Both the social function and the symbolic meaning of the festival are closely related to a series of overt values that the community recognises as essential to its ideology and worldview, to its social identity, its historical continuity and to its physical survival, which is ultimately what the festival celebrates.

Falassi's (1987, 2) consideration of the close affiliation of festivals with a "series of overt values" of the celebrating community elucidates how festive events produce meaning. In the context of the WOW festival in Hull, the sociocultural practice of meaning-making is expressed in the celebration of gender equality as a value acknowledged and highlighted through the event.

As Falassi's (1987) and Finkel et al.'s (2013) attention to the practice of celebrations outlines, festivals do not take place in a vacuum. The scholars highlight the relevance of negotiations of values through celebrations as expressions of their contemporary situatedness. Events reproduce social relationships, conversations, engagements or – more simply – culture. The global network of WOW festivals responds discursively to the particular situatedness of each host city, as

124 *Barbara Grabher*

WOW Hull illustrates. The event, and its programme of speakers, performances, and interventions, produces a particular vision of equality situated in the locality and temporality of the celebrations. In the awarding of the trailblazer status, WOW Hull explicitly highlighted local women and their achievements as an expression of the locality. Therefore, the celebration of WOW Hull reproduces and produces conceptualisations and considerations of what equality can mean in Hull in 2017. Consequently, WOW Hull illustrates how celebrations are situated within a reality that ultimately shapes their inherent content.

With awareness of the potential of meaning-making through celebrations on a discursive level, my analytical focus shifts towards the infrastructures enabling the celebration of equality. Equality-themed events embed within the wider festival landscape and urgently need to address the inequalities existent and documented in festival grounds and the festival industries. Contemporary discussions show that festivals reproduce structural inequalities and injustices in their infrastructural framework. Gisbert and Rius-Ulldemolins (2019) highlight that due to the emphasis on subversive, transgressive festival experiences, there is often a failure to understand how hegemonic social and cultural structures and controls still govern these festival spaces. The authors' call for attention is answered by increasing academic and non-academic discussions of inequalities in festival contexts. Unequal representations in line-ups, gender-based violence and exclusionary access policies are just a few examples discussed in the studies of festivals. In their considerations, Gisbert and Rius-Ulldemolins (2019), as well as other authors (see Bows, 2019; Cvetkovich & Wahng, 2001; Eder et al., 1995; Fileborn & Wadds, 2019; Fileborn et al., 2018; ND, 2017; Papisova, 2018; Staggenborg et al., 2018; Thomas, 2017), highlighted the inequalities, injustices and violence that is inherent to the infrastructures of celebrations. Traditionally characterised as a "time out of time" (Falassi, 1987, 2) in a "place out of place" (Turner, 1987, 76) and associated with experiences of norms out of the norm, festivals, through their infrastructures, need to be questioned in relation to their perpetuation of structural discrimination, which shapes society and, therefore, the celebrating community.

From a conceptual point of view, the analysis of celebrations of equality takes place on discursive and structural levels. Festivals are characterised through their potential for meaning-making as their space-time contribute to the negotiations of values. However, as embedded within the temporality of their own situatedness, festivals cannot be reduced to liminal, transgressive spaces subverting pre-existing norms. This becomes prevalent in consideration of the infrastructural conditions and in the analysis of structural violence and injustice inherent in such spaces. When festivals claim equality in their discursive outlines, the infrastructures contribute essentially to such celebrations of the sociocultural values as meanings are being made in plaster forms. As previously outlined, I am not scrutinising the relevance of celebrations of equality in the form of festivals or the particular WOW festival in Hull. But my intention is to argue that festivals – even those with a focus on equality – need to be read within the sociocultural conditions in which they are embedded. Structural injustice runs through society, and, therefore, affects the context of celebrations, as I illustrate in the following analysis.

Infrastructures of in/equality

In the conceptual discussion of festivals, I argued that celebrations do not happen in a vacuum; rather, they need to be considered in the sociocultural contexts in which they are celebrated. Independently of the focussed content of the event, festival frameworks embed in pre-existing structures. In acclaimed celebrations of equality, events risk their own values, as forms and content might not coalesce.

In the context of the WOW festival, I illustrate in the following analysis how such contradictions play out. While this was not the main focus of the interviews held with producers, artists and visitors, research participants continually remarked on elements and encounters of structural injustice within the framework of the event. I have clustered these issues in the arbitrary system of the micro-, meso- and macro-structures of inequalities. As explained in further detail next, the micro, meso and macro considerations of structural discrimination in the WOW festival in Hull refer to personal and even intimate experiences but also expand to collective concerns and overarching structures of inequalities in the event.

Perceiving in/equality

The experiences of observing-participant Sophia invite me to consider the very personal and even intimate encounters of structural inequality in WOW Hull. Sophia's experiences refer strongly to expectations and anticipations regarding the event. Her narrations tell of a journey of multiple experiences that need to be regarded in their totality in order to understand the structural dispute and their effects.

Sophia is in her mid-40s and identifies as a woman. As referred to in the following quote, she has experienced the transition from one gender to another and identifies with the label 'trans'. Sophia joins the observation opportunity enthusiastically and does not mention any concerns with her participation to me prior to the observations. It is only towards the end of the follow-up conversation, in the week after the festival, that Sophia shares with me her fears and doubts, which accompanied her prior to any WOW related events. She explained,

> (Deep Breath) Right. At the very beginning of the WOW Think-ins and then again at the festival, I was very worried and concerned. Right until I stepped out the door and went to the launch, I was very worried that I would not be able to feel part of the event because of not being born female. [...] I was very concerned that in some way, I would feel like an outsider and there would be a sort of an underlying thing, "oh, why is she here? She does not have the same issues that we have". So, yeah, whether I would be accepted into the space [was a concern of mine]. That space being the festival, you know.
> (Sophia, Observing-Participant, 15.3.17)

On the basis of her identification as "not born female", Sophia experienced difficulties associating herself with the event. She relates her disassociation with the content of the festival as expressed in the titles, labels and programme. The explicit wording of 'women' in the title as well as the programming spectrum of the festival triggered Sophia's concerns. She explained, "When we looked at the programme, there was nothing that referred to Trans and there was nothing that referred to lesbian women. Nothing at all. There was no mentioning of sort of queer politics. [...] All that was missing" (Sophia, Observing-Participant, 15.3.17). The lack of acknowledgement of non-normative forms of femininity re-enforced Sophia's doubts.

One might argue that Sophia's experience is a subjective and singular perception due to her own preconceptions. However, I counter such simplified declarations in support of the experiences of a marginalised identity within a mainstream event. I read Sophia's considerations as a call for structural disputes. The event organisers did not make an explicit effort to clarify the notions and labels used for the event. Sophia clearly remarks that the enthusiasm for her attendance was motivated by her role as an observing-participant and that the festival's accessibility felt restricted by the anxieties caused by a lack of definition and acknowledgement of marginalised identities at the event celebrating gender equality.

Her initial concerns, fears and doubts of being or feeling "like an outsider" at the event were countered by the actual experience of attending. Sophia reflects,

> [Once I was at the festival,] I made absolutely sure that I would have my voice heard. That was my way of being visible, of saying that I had something to say about the issue and that my voice was just as valid as all of the other women that were in the room. I was concerned I wouldn't feel welcomed but I seemed to be welcomed quite a bit. [...] I surprised myself by getting through it and I was surprised by the response of people to me and so [pause] I am glad I went, because, in some small way, it just justified my place on the planet and I felt part of society – rather than always feeling like an outsider.
>
> (Sophia, Observing-Participants, 15.3.17)

The positive experience of her attendance does not diminish the previous structural struggle that Sophia narrated. While marked by a 'happy ending', Sophia's experience of structural violence on the micro-level of her personal position within the event demands attention in an event celebrating gender equality.

Ticketing in/equality

Next to the micro-scale affecting the very personal encounter of structural violence, research participants urge me to look at a meso-level of structural disputes. Hereby, the experiences of structural injustice are still felt on a personal level but engages a wider collective of audiences as visible structures restrict experiences of equality.

The materiality of access – ticketing and spatial arrangements – caused emotional debates about infrastructures of inequality inherent in celebrations of equality. Observing-participant Rachel synthesises the debate well as she explains, "I was pretty angry [and thought I had to] explode at some time, because people like me [...] can't spend or don't have the spare cash to spend 10 or 15 pounds to go to a festival" (Rachel, Observing-Participant, 22.3.17).

Rachel's emotional reaction to the ticketing circumstances is informed by this lived reality in her community. Rachel is in her late-40s and identifies as a woman. Her motivation to join the research project derives from her hope to "get out to mingle more" (Rachel, Observing-Participant, 22.3.17). As a stay-at-home mother with a teenage daughter, the costs of leisure are a major concern for her. While she considers herself to be "quite well off" (Rachel, Observing-Participant, 22.3.17), she is aware and considerate of members of her neighbourhood community who are struggling with high levels of deprivation.

The logistics and pricing were a major concern for the majority of observing-participants. Day passes were priced at 10 GBP per day. The opening and closing events were additionally ticketed and required extra payment. As Rachel points out, the ticketing structure required potential visitors to have 10 GBP spare and available at the moment that ticket sales went live. Additionally, a quarter of events in the festival required additional registrations. While I completed such registrations for all observing-participants in order to facilitate their attendance at the festival, I experienced difficulties registering for appropriate time slots and shows due to the complex process. While ticketing mechanisms are a form of crowd management, in the context of a celebration of equality and in the socio-economic realities of a city like Hull, festival tickets are an infrastructure of inequality, excluding, particularly, women and marginalised groups in the population. Therefore, ticketing is a prime example of structural disputes in the context of the WOW festival in Hull.

Furthermore, meso-levels of structural disputes also include considerations of spatial arrangements of festivals. While festival venues are continuously under discussion for their contribution to in/equalities, my attention is on the spatial arrangements within festival venues, and, particularly, it is on the provision of a VIP space and treatment, which suggests an immediate structural dispute. I draw on my own observations for this discussion. As I supported the WOW Circle of Friends, I received free tickets to the opening ceremony for the festival. On the evening of the opening, while around 40 people waited on the main plaza in front of city hall, I walked into the main foyer to pick up my tickets and was asked by a volunteer to follow her. Without having received any information about special treatment, I was taken to a back room behind the main stage, where I encountered many WOW Circle of Friends members, performers in the festivals and the Hull2017 Ltd production team enjoying a free buffet of canapés and wine. In a short speech, chair of the board of Hull2017 Ltd, Rosie Millard, invited us to the opening ceremony and expressed her gratitude

128 *Barbara Grabher*

on behalf of the team of Hull2017 Ltd. Sophia, who was with me at the time, reflected on the arrangement:

> Glamour Glamour Glamour. All of the sort of knows and faces in the city were there [in this VIP space], you know, all of the 'best' people were there. They had that VIP lounge, didn't they? For all the special people, the business leaders and the special people in the city, people, who sort of exist in the top layer of the pile and not the regular folk, who had to wait outside.
> (Sophia, Observing-Participants, 15.3.17)

While I understand that this treatment and the spatial segregation in the form of a VIP lounge derives from a grateful intention, as a member of the production team explained to me later, the mechanism contrasts with the ethos of the festival.

Marketing in/equality

The final layer of encounters of structural disputes addresses the overall framework from which the WOW festival derives and in which it is placed. In this consideration of the macro-structures, I draw upon statements from observing-participant Daniel. Daniel is in his mid-30s and identifies as a man. He joined the team of observing-participants as a form of training for himself. As a musician and music teacher, he actively seeks out events relating to the performing arts in order to engage with new ideas and the associated community in Hull. Therefore, in his visit to the WOW festival, he focussed mainly on the performances taking place during the event and reflected in detail about their narratives. Beyond this focus, when questioned about how the festival contributes to cultures of equality in the city, Daniel expressed strong views in opposition to the idea that an event like WOW Hull can be true to its celebrated values. He synthesised,

> No festival like this one can achieve a culture of equality because they all base in capitalist principles of profit. It is like there is more marketing than content. Around the world, all the big festivals work in this way, and it is an economic question, not a cultural one.
> (Daniel, Observing-Participant, 15.3.17)

Daniel's considerations call attention to the overarching structures of how the event is produced and promoted. As a global festival movement for gender equality, Southbank Centre London constructed a brand, which travels similarly to a franchise system around the world. The festival's identity is defined by the logo, which marks the festival heavily through its colour scheme of red, yellow and black. Furthermore, the structures of programming through community brainstorming called WOW Think-ins conditions the celebrations. Even within the festival programming structures, certain elements are borrowed from the original festival concept. In 2019, the WOW Foundation established itself as a company to run the London event, as well as the franchise of the festival brand. The exact conditions for festival franchisees are not made public in the official web presentation of the WOW Foundation or the WOW Hull presentation. However, I estimate that the co-operative perspectives as addressed by Daniel have strongly influenced the development of the event and its expansion. Therefore, on a macro-level of observed structural disputes, Daniel's observation encouraged me to ask how equality, a celebrated value, becomes a commodity in the context of a branded festival. While I refrain from an oppositional binary and rather refer to a mutual co-existence of profit-focussed capitalist intentions and the celebration of equality, this reality of structural injustice needs to be examined in further detail through future research.

Discussion and conclusion: the fragile notion of equality

Drawing on producers', artists' and visitors' perspectives of the WOW festival in Hull in 2017, this chapter discussed structural disputes which are experienced in the celebration of gender equality. By highlighting examples of perceived structural inequalities on a micro-, meso- and macro-level, I engaged with a nuanced understanding of what festivals and events can do. While often discussed in terms of a liberation due to the liminal experiences with which festivals are associated, the discussion of structural violence allowed me to shine a light on the variety of aspects that merge in the complexity of the festival environment. The WOW festival in Hull produces and reproduces in its content notions of equality, which are crucially embedded in the local society. However, when investigating the form and framework of the festival, structures and conditions of inequality appear, which put the celebrated value at risk. In reference to Sophia's experience, I highlighted the very personal concern of accessibility as the label of 'women' was neither challenged in the programme outline nor clarified in any other form. On a meso-level of collective encounters with structural inequalities, I addressed Rachel's consideration of the ticketing prices as systems of exclusion. Additionally, I highlighted concerns about the spatial arrangement in respect to my own observations and questioned how the establishment of a VIP space limited considerations of equality in the celebration. Finally, through Daniel's remark on the commodification of equality, I suggest that the franchise structures, in which the festival brand WOW is embedded, require further attention.

The analytical examples contributed by Sophia's, Rachel's, Daniel's and my own observations do not scrutinise the WOW festival and its ambition to celebrate equality. However, conceptually and analytically, the observations invite considerations of the fragility of equality. While clearly embraced in the content of the festival, the practices and the processes of the celebration need to be embedded in structures of equality in order to support the celebrated value. Festivals and events hold productive and reproductive capacities in society. The celebration of equality in the form of the WOW festival represents a great achievement and important development in consideration of the struggle for gender equality. However, continuous attention to the holistic practices of equality is crucial in order to avoid structural disputes between content on equalities and forms of inequalities.

Acknowledgement

This project received funding from the European Union's Horizon 2020 research and innovation programme under the Marie Skłodowska-Curie grant agreement No 675378. My gratitude goes out to all research participants who continuously support the investigation and challenge, reflect on and discuss the developments in and of their city with me. Special thanks go to Dr Tommaso Trillo and Dr Jennifer Jones for their feedback and proofreading.

Note

1 Mary Wollstonecraft was an eighteenth-century writer, philosopher and feminist activist who lived in the market town of Beverley, which is close to Hull (HU17 Admin, 2018). The Headscarf Revolutionaries were a Hull-based group of women who campaigned for safety regulations on the trawler fishing boats in the 1960s and 1970s (Lavery, 2015). Maureen Lipman, CBE, is a film, theatre, and television actor who was born in Hull (ND, n. d).

References

Bows, H. (2019). How can sexual assaults at festivals be stopped? *BBC News*, June 7.
Culture Place and Policy Institute. (2018). *Cultural transformations: The impacts of Hull UK city of culture 2017*. Hull: University of Hull.
Cvetkovich, A., & Wahng, S. (2001). Don't stop the music: Roundtable discussion with workers from the Michigan Womyn's Music Festival. *Journal of Lesbian and Gay Studies*, 7(1), 131–151.
Dashper, K. (2016). Researching from the inside: Autoethnography and Critical Event Studies. In R. Lamond & L. Platt (Eds.), *Critical event studies: Approaches to research* (pp. 213–231). London: Macmillan Publishers Ltd.
Department for Culture Media and Sports. (2013). *UK city of culture 2017: Guidance for bidding cities*. London: Department of Culture Media and Sports.
Eder, D., Staggenborg, S., & Sudderth, L. (1995). The national women's music festival: Collective identity and diversity in a lesbian-feminist community. *Journal of Cotemporary Ethnography*, 23(4), 485–515. https://doi.org/10.1177/016344300022005001
Falassi, A. (1987). Festival: Definition and morphology. In A. Falassi (Ed.), *Time out of time: Essays on the festival* (pp. 1–13). Albuquerque: University of New Mexico Press.
Fileborn, B., & Wadds, P. (2019). How music festivals can change the tune on sexual violence. *The Conversation*, January 9, pp. 1–5.
Fileborn, B., Wadds, P., & Tomsen, S. (2018). New research shines light on sexual violence at Australian music festivals. *The Conversation*, October 23.
Finkel, R. (2009). A picture of the contemporary combined arts festival landscape. *Cultural Trends*, 18(1), 3–21. https://doi.org/10.1080/09548960802651195
Finkel, R. (2015). Introduction to special issue on social justice & events-related policy. *Journal of Policy Research in Tourism, Leisure and Events*, 7(3), 217–219. https://doi.org/10.1080/19407963.2014.995905

Finkel, R., McGillivray, D., McPherson, G., & Robinson, P. (2013). Introduction. In R. Finkel, D. McGillivray, G. McPherson, & P. Robinson (Eds.), *Research themes for events* (pp. 1–6). https://doi.org/10.1079/9781780642529.0000
Finkel, R., & Sang, K. (2016). Participatory research: Case study of a community event. In R. Lamond & L. Platt (Eds.), *Critical event studies: Approaches to research* (pp. 195–213). London: Macmillan Publishers Ltd.
Gisbert, V., & Rius-Ulldemolins, J. (2019). Women's bodies in festivity spaces: Feminist resistance to gender violence at traditional celebrations. *Social Identities, 25*(6), 1–18. https://doi.org/10.1080/13504630.2019.1610376
Grabher, B. (2018). Observing through participants: The analytical and practical potential of citizens' involvement in event research. *Studies on Home and Community Science, 11*(2), 66–76. https://doi.org/10.1080/09737189.2017.1420389
Harding, S. (1993). Rethinking standpoint epistemology: What is "strong objectivity"? In L. Alcoff & E. Potter (Eds.), *Feminist epistemologies* (pp. 49–82). London and New York: Routledge.
Hilgers, M., & Mangez, E. (2015). Introduction to Pierre Bourdieu's theory of social fields. In M. Hilgers & E. Mangez (Eds.), *Bourdieu's theory of social fields: Concepts and applications* (pp. 1–36). London and New York: Routledge.
Hirsch, E., & Olson, G. (1995). Starting from marginalized lives: A conversation with Sandra Harding. *JAC: A Journal of Composition Theory, 15*(2), 193–225. https://doi.org/www.jstor.org/stable/20866024
HU17 Admin. (2018). The first feminist: Mary Wollstonecraft (1759–1797). Retrieved from http://www.hu17.net/2018/08/28/the-first-feminist-mary-wollstonecraft-1759-1797/
Hull 2017 Ltd. (2015). *Hull UK city of culture 2017: Strategic business plan 2015–2018*. Hull: Hull 2017 Ltd.
Hull 2017 Ltd. (2017a). *Made in Hull: Season guide Jan-Mar*. Hull: Hull 2017 Ltd.
Hull 2017 Ltd. (2017b). *Roots & Routes: Season guide Apr-June: Freedom – season guide Jul-Sep*. Hull: Hull 2017 Ltd.
Hull 2017 Ltd. (2017c). *Tell the world: Season guide Oct-Jan*. Hull: Hull 2017 Ltd.
Hull 2017 Ltd. (2017d). *WOW Hull (Women of the world)*. Retrieved from www.hull2017.co.uk/whatson/events/wow/
Hull City Council. (n.d.). *History of Hull*. Retrieved from www.hullcc.gov.uk/portal/page?_pageid=221,148379&_dad=portal&_schema=PORTAL#
Lamond, I., & Platt, L. (2016). Introduction. In I. Lamond & L. Platt (Eds.), *Critical event studies: Approaches to research* (pp. 1–15). London: Palgrave Macmillan.
Lavery, B. (2015). *The headscarf revolution: Lillian bilocca and the triple trawler disaster*. London: Barbican Press.
Mayring, P. (1994). Qualitative inhaltsanalyse. In A. Boehm, A. Mengel, T. Muhr, & Gesellschaft fuer Angewandte Informationswissenschaften (Eds.), *Texte verstehen: Konzepte, Methoden, Werkzeuge* (pp. 159–175). Konstanz: UVK Univ.-Verl. Konstanz.
N.D. (n.d.). *Maureen Lipman: Biography*. Retrieved from www.imdb.com/name/nm0513520/bio
N.D. (2017). Swedish music festival is cancelled in 2018 after rape and sexual assault claims. *BBC News*, July 3.
Office for National Statistics. (2019). *Population estimates for UK, England and Wales, Scotland and Northern Ireland: Mid-2018, using April 2019 local authority district codes*. Retrieved from www.ons.gov.uk/peoplepopulationandcommunity/populationandmigration/populationestimates/datasets/populationestimatesforukenglandandwalesscotlandandnorthernireland

Papisova, V. (2018). Sexual harassment was rampant at Coachella 2018. *TeenVogue*, April 18.

Pavoni, A., & Citroni, S. (2016). An ethnographic approach to the taking place of the event. In R. Lamond & L. Platt (Eds.), *Critical event studies: Approaches to research* (pp. 231–253). London: Macmillan Publishers Ltd.

Redmond, P. (2009). *UK city of culture: Vision statement*. London: Department for Culture, Media and Sports.

Southbank Centre. (2018). *WOW: Women of the world*. Retrieved from www.southbankcentre.co.uk/whats-on/festivals-series/women-of-the-world

Staggenborg, S., Eder, D., & Sudderth, L. (2018). Women's culture and social change: Evidence from the national women's music festival. *Berkeley Journal of Sociology*, *38*(1993–1994), 31–56. https://doi.org/www.jstor.org/stable/41035465

Thomas, R. (2017). A remarkable absence of women: A comment on the formation of the new events industry board. *Journal of Policy Research in Tourism, Leisure and Events*, *9*(2), 201–204. https://doi.org/10.1080/19407963.2016.1208189

Turner, V. (1987). Carnival, ritual and play in Rio de Janeiro. In A. Falassi (Ed.), *Time out of time: Essays on the festival* (pp. 74–90). Albuquerque: University of New Mexico Press.

10 The green wave!
Resistance in festivals as part of the women empowerment process in Argentina

Valeria Vegh Weis and Lucía Montenegro

Introduction

Robbed, assaulted, harassed. Until now. ... The chapter explores the case of Argentina where an impressive feminist movement has been aroused in the last few years. The movement encompasses the 'green wave', a growing claim for legal, free and safe abortions, as represented by a green handkerchief that contrasts the sky-blue one of those known as 'pro-life',[1] and other demands such as the "Ni una menos" (Not even one less – concerning femicides).

Within this resistance movement, some voices started to call attention to the violence, inequality and sexism that rule in festivals and concerts. Before the 'green wave', Argentinian women[2] had been mostly suffering these burdens alone as the price to be paid for entering a masculine space. For example, attending a concert seemed indistinguishable from the experience of being, at least, unwillingly touched when immersed in the crowd; harassment was implicitly accepted as an extra-cost attached to the entrance to a packed public event. However, the 'green wave' allowed women to acknowledge the collective character of these experiences. For example, a poll carried out in 2018 showed how widespread the verbal sexual harassment in the streets (Instituto de Proyección Ciudadana de la Ciudad de Buenos Aires, 2018) was and, with this, the collective character of the unpleasant experience. It was then that many harmful behaviours and experiences suffered by women individually as unquestionable burdens started to be called out.

As a result of this breaking point, resistance began. Particularly within festivals and music-related spaces, it is possible to individualise three big trends of feminist resistance. First, different women organised themselves to publicly shame and even present criminal charges against musicians who had sexually assaulted them. Second, women activists started to organise 'safe events' where only women can attend. Third, women-only music festivals were organised for the first time.

Following McGillivray and Jones (2013), the chapter explores these trends, focussing on the new methods of coordinating and organising resistance as events in themselves. In the Argentinean case, we can see resistance as a double event: as the coordination of efforts against a different kind of gender oppression (resistance as an *event*) and as the performance of specific cultural activities (*events* in the traditional sense).

Notably, as most of these experiences are happening now and in the underground scene, they are not (still) covered by the academic literature or mainstream media. Thus the sources used in the chapter mainly come from alternative magazines and even Facebook events.

Settled on a critical perspective, the suggested study understands that it is indispensable to draw a connection amongst harms involving women in festivals and music-related spaces, the bias of the socio-economic-cultural context in which these phenomena take place and the resistance strategies that the country is exposing. This critical approach, it is argued, needs to also be enhanced by a southern perspective to highlight the particularities of harms and strategies involving people from the Global South.

Resistance as a (necessary) event and the particularities of the Global South

It is important to refer to Butler (1993, 1999) when she shows how the subject position of women comes into being, is regulated and can be governed. This happens in the workplace, in the family, in the streets and, also, in music festivals and the music industry in general. This is why it is urgent to "contest the widely-held notion that festivals are temporal spaces free from structural sexism, inequalities, or gender power dynamics" (Platt and Finkel, 2020). In contrast, festivals are spaces where those modalities of oppression are enhanced because the environment serves as an excuse or as an opportunity to victim blame and shame.

Following this, if alternative discourses and narratives are going to be carried out to secure social change, they will have to occupy those conflicting spaces. Indeed, as McGillivray and Jones (2013, 130) clarified, "Leisure practices could be employed effectively to develop oppositional identities for the disenfranchised and dis-empowered masses". In the same vein, de Jong (2017) and Sharpe (2008) referred to the use of festivity to engender social change in relation to queer gatherings and music festivals with political aims, respectively. Even more radical, Rojek (2005) suggested that alternative or oppositional spaces can even challenge the rule of capital and the state.

Feminist approaches to festivals and music-related events can contribute to clarifying that the public space is not neutral at all; instead, it is a gendered, sexualised and racialised arena – a site for the maintenance and reproduction of complex power relations (Scraton & Watson, 1998). Moreover, a feminist perspective can help conceive festivals as new spaces that are "neither the masculine public nor the feminised private" (Rose, 1993, 136). This means that a feminist perspective has the potential to not only approach the festival as an area that can be known and mapped, as masculine views suggest (ibid.), but also as a complex space that includes emotions and collective experiences.

Following this logic, and from an intersectionality perspective, women's experiences are conceived as traversed by racial, class, ethnic and age factors, amongst others. Here we would like to draw attention to the geographical fact, as the experience of the Global North and the Global South can be diametrically different. Indeed, while most of the northern countries have legalised abortion decades ago, this is still a pending demand in the Global South. Including the experiences and perspectives of the Global South might foster the democratisation of knowledge, help to exchange unique experiences happening in one and another part of the globe and open the spectrum of the phenomenon under study. Specifically, in this case, a southern perspective can shed light on the relationship between the socio-economic situation in Argentina and the resistance strategies carried out by the women's movement. For example, activists not only have to guarantee 'safe spaces'; they have to do so in an affordable manner to avoid excluding vast parts of the population suffering from sexism and economic difficulties.

Importantly, the distinction of North-South refers to the division between Western Europe and North America on the one side and Latin-America, Africa, Asia and Australia on the other (Carrington et al., 2018). However, it is possible to suggest that this distinction might be too simplistic, as Australia, China or Japan, to name a few, are pioneers in the development of knowledge on a global scale (ibid.). Moreover, we should also understand this North-South relation within states (Vegh Weis, forthcoming). In this logic, it might not even be enough to incorporate the experience of a southern country such as Argentina to expose a southern perspective if we do not also acknowledge the different realities within the country. As we will see, most of the experiences exposed in this chapter happen in the three most important cities (Buenos Aires, Córdoba and Rosario), and more work needs to be done to expose the reality in less central regions. Indeed, worldwide, knowledge tends to be concentrated in the capital cities and certain institutions (representing the geographic north within the country). Moreover, these institutions are usually ruled by individuals with a good position in terms of class, gender, ethnicity, religion and age (representing the demographic north). Going back to the example of Argentina, most of the books, articles, conferences and syllabus are designed by old white men working at established institutions in the capital city. Only recently have networks of women professors at some law schools started to gather and claim a role in the academia.[3] Notably, at Buenos Aires University, the network of women law professors was inspired when the law school held a panel on obstetric violence and abortion – an indisputable women's issue – that was only led by men. Additionally, other minorities, such as indigenous groups, ethnic minorities or LGTBQ+ communities, are still far from being recognised as legitimate producers of knowledge (ibid.).

Overall, we could understand South, first, as a geographic metaphor that involves the former colonies that remain low-income or middle-income countries without a key place in world politics and that, therefore, lack a central role in the global production of knowledge. Second, we could also understand South as a geographic metaphor of those places within each region, country, city or location that are less developed and have a lower impact on decision-making processes at the socio-economic and political levels, all of which makes them less relevant in terms of knowledge production. Third, we could also understand South as a demographic metaphor of those whose voices are not usually heard and whose problematics, experiences and perspectives do not appear in mainstream theories (ibid.).

Importantly, none of this should imply a normative statement: knowledge from the South (understood in these two geographical levels and the demographic one) is not necessarily better than the North's or even good at all. It is just different, in some cases, and it should be heard if we aspire to build a truly global theory of criminological knowledge attentive to different problems and perspectives (ibid.).

The context of the women's movement in Argentina: "not even one less" and the green wave in Argentina

In 2010, Wanda Taddei was burnt alive by her partner, a musician in the rock band Callejeros. This triggered a 'contagious effect', with 136 cases of women burnt alive by their partners between 2010 and 2012 (Página/12, 2014). Some state that the Wanda's case produced an increase of femicides. Others state that the number of femicides remained the same and that what increased was the number of denounces of cases that were taking place without being noticed. In 2012, as a result of the public effect of the femicides, Argentina passed a specific criminal statute aggravating the punishment for homicides against a woman because of her condition (Infoleg, 2019). The law did not stop the death count.

In May of 2015, the body of Chiara Páez, a 14-year-old girl from Santa Fe, was found buried in her boyfriend's backyard. Chiara was pregnant at the time. This was the breaking point for the emergence of the so-called "Not Even One Less" movement to stop femicides in the country for real. More than 300,000 people gathered at Congress Square (La Nación, 2018). Notably, the "Not Even One Less" claim was made by the Mexican, feminist poet Susana Chavez, killed in Ciudad Juarez in 2011: "Not even one woman less, not even one more dead woman" was her usual statement in the demonstrations (Página/12, 2015). The movement in Argentina has been growing since then, gathering demands for gender equality, legal abortion and the end of femicides (Ni Una Menos, 2017).

Particularly relevant is the role of the Campaign for the Legal, Safe and Free Abortion (Campaña por el Aborto Legal, Seguro y Gratuito, 2019) that spread information on comprehensive education, contraceptive methods and bill projects to legalise abortion. It started in 2005, but the campaign only became massive recently along with the "Not Even One Less" movement. In 2018, the members of the campaign presented the bill to legalise abortion before the Congress as they do every year, but, on this occasion, the public response was massive. There were two weeks of public hearings with almost a thousand speakers raising their arguments in favour and against the project, and the mobilisation in the street was unprecedented (Timerman, 2018). The bill did not make it to the

Senate but was an opportunity to strengthen the feminist movement via public events.

Resistance trends in festivals in Argentina

Violence against women in the music industry and music festivals

The patriarchal preconception takes it for granted that if a young woman admires a member of a band, she would be delighted to be approached and seduced by him. Informed consent is not regarded as necessary. It is supposed to be a dream come true. However, in Argentina, many women do not agree with these statements anymore. They became aware that they had only consented to attend a show or get to know their admired musicians personally but not to be touched without being asked. That was not a dream come true; it was harassment and sexual assault.

Within the inspiration of the "Not even one less movement" and the feeling of sisterhood developed amongst women everywhere in the country, many women decided to claim "Stop!" Indeed, many of those who have experienced sexual abuse by male musicians when attending their shows or afterwards started to gather. They created the group "We Will Not Remain Silent Anymore" (Spanish, Ya No Nos Callamos Más) in November of 2016, which, since then, served as a space to report abuse and other gender-related violent acts in the music business and at music festivals. It is an independent and nonpartisan space – a safe space where women can build ties, share their experiences and seek justice. Their blog claims, "We are looking forward to deconstructing this system, moving out of our comfort zones and creating something new" (Ya no nos callamos más, 2019).

Indeed, as Ileana Arduino, a feminist researcher and attorney, claimed,

> As we become conscious about the abuses, subjugation, and violence we feel urged to take action. It's a horrible experience when we realize the levels of aggression that we sublimated and hidden and how much we have been used to tolerate and remain silent. . . . Now, we are all collectively challenging these situations, renaming formerly naturalized practices as violence legitimized by the patriarchal culture and pointing out the existing gender hierarchy.
>
> (Arduino, 2018)

The first case carried out within this resistance reaction was against Cristian Aldana, frontman and singer of the alternative rock band El Otro Yo and then president of the Independent Musicians Union. Victims started to share their testimonies, leading to the first massive report of sexual crimes in the local music world: seven victims who were between 13 and 18 years old when the abuses took place took the lead and brought the case to court. Like many other musicians, the court stated that Aldana had been taking advantage of his fame and idol status to invite his fans to parties and abuse them, and it convicted him to serve 22 years in prison.

However, the victims do not look for retribution as the main goal. As Arduino (2017) claimed, if the feminist movement is trying to change the structural nature of violence and demand radical transformations, it cannot do so through punitive solutions that are only individually based, enhance violence and do not question the system as a whole. Instead, most of them are more committed to non-repetition. As Carolina Luján, one of Aldana's victims, stated in an interview, "We must draw attention to this [kind of event], that we talk about it. ... We have to assume this responsibility, so it won't happen again" (Lahiteau, 2018). Moreover, this experience is also looking to build a more permanent safe space where women can be heard without suspicion. In this logic, they are trying to bring legitimation to the victims' testimonies, which are usually undermined by the musicians' attorneys and mainstreaming media (Lahiteau, 2018).

Other exposed cases of rape and harassment involve José Miguel "Migue" Pópolo, frontman of the indie band La Ola Que Quería Ser Chau. One of his victims uploaded a video recounting that she was raped and, since then, two other victims were inspired to denounce him. Pópolo denied the accusations (Lahiteau, 2018) but he is in pretrial detention for rape (Rolling Stone, 2018). Walas, singer for the hard-rock band Massacre, made fun of the victim's video, saying, "In the video, she [the victim] says that she was raped twice. ... What did they do between rapes? Did they smoke a cigarette?" There was such an outcry, and the crowd became so wild about this statement, that Walas had to apologise in public. He also apologised on Facebook (Silencio, 2016).

Following the same logic, a girl of the northern province of Jujuy accused Juan Sebastián "Juanse" Gutiérrez, the frontman of the famous rock band Ratones Paranoicos, of gang raping her in a van with six other members from another music group (Lahiteau, 2018). The last available information is that a criminal case was presented in Jujuy (El Tribuno, 2014a, 2014b; La Voz, 2014). The musician responded with insults (La Voz, 2015), but he apologised days later after a demand was presented before the National Institute against Discrimination, Xenophobia and Racism (El Tribuno, 2015). In another case, several testimonies against Santiago Aysine, the singer for Salta La Banca, became public (Clarín, 2017). Aysine also denied the accusations at first, but afterwards, he acknowledged that he used to be violent and sexist: "I made mistakes, I am beginning to understand everything I did wrong and I am sorry", he stated while announcing that he was leaving the band (Clarín, 2017). Fidel Nadal, the leader of the band Todos tus Muertos, was also accused by his former girlfriend of verbal, sexual and symbolic violence and abuse (Lahiteau, 2018). In 2018, the victim presented a civil claim, and a restriction order was issued (La Nación, 2018). During a public lecture in a journalism institute, Gustavo Cordera, a popular rockstar, former frontman of the rock band Bersuit Vergarabat and now solo artist, said that "there are women who need to be raped because they are hysterical" (Fiscales, 2018). The musician faced a large public condemnation, mostly in social media, but also at his shows. Moreover, an analysis of his lyrics and their underlying violence and vindication of the rape culture was exposed (Arduino, 2017). Cordera had to face trial and do two benefit concerts, apologise publicly and attend two workshops on gender violence (Fiscales, 2019).

In October 2018, a group of 40 women presented a demand against members of the rock band Onda Vaga for mistreatment and sexual and psychological abuse (Página/12, 2018). In November, the band issued a communication acknowledging that they had been insensitive but that they were not criminals or sexual offenders (Infobae, 2018). The 40 women were harassed by the media and ended up shutting down their blog (La Nación, 2018). In turn, the singer for the rock band Los Espíritus was denounced for sexual abuse, and the band decided to ask him to leave the group, but they overturned their decision weeks later. The singer apologised in public and affirmed that he was not the same as he was 11 years ago when the events happened (Tiempo Argentino, 2019; La Nación, 2018).

These are only a few examples that expose the general outline of a music industry based on a chauvinistic culture that hides and legitimates abuse and violence against women. But they are also examples of resistance. Since 2016, the feminist wave allowed those who had not been able to raise their voices to finally do it. One testimony has been enough to inspire other women to come forward and share their stories to change the patriarchal and misogynistic culture that reigns in the Argentinean music industry and music festivals. Because it exposes that other types of festivals are possible.

Safe events

Women suffer from oppression, sexual, physical and psychological violence and socio-economic inequality. It would be discouraging to carry out a struggle that is also based on suffering and pain. That is the philosophy behind *Marilina Tortillera*, a lesbian-run, girl-only party named after Marilina Ross, an Argentinian actress, musician and artist who wrote one of the local LGTBQI+ anthems.

Using a US contemporary feminism framework, the party can be framed as a 'safe space', even though this is not a widespread notion in Argentina. The party was born in 2016 in a "hostile context, where inflation was increasing and the value of the salary was decreasing day after day, while tickets for parties were becoming more and more expensive", as Lía, one of the organisers described (Caja de Herramientas, 2018). To confront one of the many downsides of the economic situation and the need to find safe spaces to strengthen women's resistance, Lía and her friends Magui and Camila brought up the idea of offering an opportunity for happiness, celebration and joy from a feminist perspective. In other words, Marilina Tortillera was born under the understanding that partying can be a form of empowerment by enjoying our bodies as they are and looking after each other in a violent-free environment where happiness can be restored. It is a party where girls can be free of the social barriers of being around boys and of the burden of sexism. As Lía explained, "We don't behave in the same way when boys are around, so we decided to build a space where girls can enjoy among themselves" (ibid.). Also, this event provides women with the opportunity to perform jobs that are usually confined to men.

But Marilina Tortillera is not only about partying; it is also a political space. It exposes the lack of safe events for women and, especially, the needs of the lesbian community. This is because, even though there are multiple gay and queer/dissident parties and events, none of them guarantees a violence-free place with a feminist perspective.

In 2019, the party celebrated its third anniversary and has become a cultural reference space for the lesbian and feminist community. Notably, on March 2019, the party was announced as "Viva La Pepa" (a local expression that means "Live the life"!) in memory of Natalia "Pepa" Gaitán, a lesbian woman from the province of Córdoba who was murdered by her girlfriend's stepfather because of her sexual orientation (Borrelli, 2019). As Lía and her friends said,

> There is no revolution without joy, nor without sad people (Caja de Herramientas, 2018). For us, to be activists in these spaces is a fundamental strategy to guarantee our pleasure. We think that the revolution has to take place in the streets, but also in our beds and in our parties. I don't know if there's anything more profoundly hurtful for patriarchy than lesbians loving each other. We think it's a political fact that every two months 400 lesbians and translesbians get together to dance and enjoy.
>
> (Borrelli, 2019)

As Goldman (1934, 56) said, "If I can't dance, I don't want to be in your revolution". With this movement, they created an event, a festival, a way of resisting without stopping dancing.

Women musicians-only festivals

Argentina has a strong music culture with festivals, concerts and events happening every day. However, women on stage are a rarity. They might be in the chorus of a rock band or offering shows for children, but it is very rare to find them as solo figures or front band members in the rock scene. At least until now.

It was recently that Marilina Bertoldi, a woman and visible lesbian artist, said in one of her concerts, "There's no lack of female bands – there's a lack of female festivals". This was enough to inspire the GRL PWR festival (girl power festival), an event aimed at disseminating the work of women musicians and calling attention to inequality in the music business, particularly in terms of female presence in festivals. Notably, not only is the musical line-up is women only but also the designers, security staff, bartenders, painters, exhibitors, producers, organisers and decorators are women. As the journalist Florencia Freijo stated, "[The festival] offers a feminist and feminine world . . . and anyone can breathe the feminist resistance" (Gentil, 2019).

The first edition of the festival took place in August of 2018 in the context of the widespread debate about abortion and women's rights in the country.

However, the struggle is not easy. In 2019, the event organisers had to publicly confront the controversial declarations of José Palazzo, the director of the most popular rock festival in Argentina, the Cosquín Rock, which takes place every year since 2001 in the province of Córdoba. Palazzo stated that the existing women artists are not talented enough to keep up with the Cosquín Rock artists and that "they do not sell tickets" (Perfil, 2019a), meaning that they do not attract the public. The representatives of the GRL PWR festival replied, "[We] will prove that women can also sell tickets" (Gordillo, 2019). Indeed, while the first edition sold out with 1,200 people in the public, this year, the number climbed up to 4,000 in its edition in Córdoba and to 1,300 in its edition in the City of Rosario, the birthplace of some of the most important and talented musicians in the country. Moreover, in this second edition, the festival obtained important sponsors, such as ConectaSkyy (of SkyyVodka). Relevantly, the organisers' response to Palazzo was also accompanied by a strong reaction in social media, and Palazzo had to apologise (Télam, 2019).

Notably, GRL PWR is only one of the expressions of this new trend that tries to enhance women's participation in cultural events. As recognition for her talent, but also as a result of the increasing role of women artists in the local scene, Marilina Bertoldi, the one inspiring these events, received the maximum Carlos Gardel Award, the greatest honour for a musician in Argentina (Infobae, 2019; Clarín, 2019). Also, a national bill establishing a 30% quota for women in every multi-band music festival was discussed and made it to the Senate (Perfil, 2019b). Interestingly, these phenomena are not separated from the national struggle for free abortions and the end of femicides. Indeed, during the GRL PWR festival, one of the artists, Erica Sativa, spoke about gender equality in music festivals, and she asked for legal, safe and free abortions (Notilife, 2019; TN, 2019). In turn, Señorita Bimbo, a comedian and journalist, highlighted the relevance of the collective nature of the movement if they aimed at achieving a political transformation:

> Our path towards empowerment is not enough, we need to do this together. . . . The green handkerchief a symbol that helps us to organize ourselves, to realize that we have something in common and that we have an opportunity.
> (Freijo, 2019)

Also, Ofelia Fernández, an Argentinian politician who is only 19 years old and emerged from the feminist movement, stated, "It's not that feminism became politicized, feminism is a political movement, we are here to create new ways of doing politics because we have a very important goal: we have to change the world" (ibid.).

Lucía Amarilla, one of the festival's organisers, added, "This event, this festival, transcends the cultural to become social and political. . . . It is a space to debate ideas and the main concerns of the feminist movement in Argentina" (ibid.). Under this logic, the festival offered not only music but also talks, twerking workshops, personal defence classes, stand-up shows, women-run food trucks and shops selling women-empowering souvenirs, fanzines and feminist literature (ibid.). GRL PWR is *resistance as an event* (denouncing the chauvinist character of the music industry became a public event) and an *event to foster resistance* (the festival as a way of advancing the role of women musicians).

Conclusions

This chapter has helped to keep building necessary literature on resistance to patriarchy and sexism as an event in itself. Moreover, the chapter showed how the feminist movement in Argentina exposes the double-character of the link between resistance and events: resistance as an event but also how certain events can foster a wider resistance movement. Focussing on the specific case of Argentina also offered the opportunity to expose experiences from a global southern country, helping to democratise the knowledge and avoiding North-minding theorisation that does not take into account what happens outside the US-European borders.

As has been shown, resistance run by women figures in the Argentinean music scene is vigorous and promising. However, it is relevant to also highlight that the two main challenges present here also have a worldwide expression, which needs to be overcome if this resistance movement wants to be as radical as it presents itself. One of them has to do with the punitive orientation of some sections of the movement. To them, resistance appears as a synonym of criminal charges and public denouncements (in the local slang, *escraches*) against those musicians accused of sexual, physical or psychological violence. Using a selective, unfair and punitive criminal justice system (Vegh Weis, 2018) cannot be the main strategy of a movement that is trying to achieve gender equality and change how things have been running. If we challenge the power of men over women by using the same instruments that they have been using to oppress us, we will end up in a vicious circle. The feminist resistance movement is powerful enough to think and develop new transformative strategies based on education, prevention and capacity building. The second challenge has to do with the exclusion of those women who do not respond to the cisgender and white model. Again, we can do more. We do not want to resist the oppression of men over women to impose the oppression of white cis-women over racially and ethnically diverse trans-women.

We have come far: let us make feminist resistance a pluri-ethnic, racial, gender, sexually diverse and non-punitive event that can change the current unfairness for good.

Notes

1 The term 'pro-life' is disputable as it presents those defending legal abortion as anti-life when what they are doing is defending the life and autonomy of the pregnant women.
2 We understand 'woman' as encompassing all people self-identified as a woman.
3 http://www.derecho.uba.ar/noticias/2018/la-red-de-profesoras-organizo-un-panuelazo-a-favor-de-la-despenalizacion-del-aborto

References

Acusan a músicos de Onda Vaga por abuso sexual. (2018b). *La Nación*, October 3. Retrieved from www.lanacion.com.ar/espectaculos/musica/denuncian-musicos-onda-vaga-abuso-sexual-nid2177994

Agenda. (2019). *Centro Cultural Matienzo*. Retrieved August 30, 2019, from http://ccmatienzo.com.ar/wp/events/marilina-tortillera-2/

Agenda Ciudad de Buenos Aires. Fiestas. (2017a). *Vuenos Airez*. Retrieved from www.vuenosairez.com/ar/ciudad-de-buenos-aires/agenda/marilina-tortillera/177241

Agenda Ciudad de Córdoba. Eventos. (2017b). *Vuenos Airez*. Retrieved from www.vuenosairez.com/ar/ciudad-de-c%C3%B3rdoba/agenda/festival-grl-pwr-2019/187214

Agenda Ciudad de Buenos Aires. Fiestas. (2018). *Vuenos Airez*. Retrieved from www.vuenosairez.com/ar/ciudad-de-buenos-aires/agenda/marilina-tortillera/183416

Arduino, I. (2017). A Cordera ni probation, a las mujeres ni justicia. *Cosecha Roja*, November 8. Retrieved from http://cosecharoja.org/cordera-ni-probation-las-mujeres-ni-justicia/

Arduino, I. (2018). No nos callamos más ¿y después? *Cosecha Roja*, April 25. Retrieved from http://cosecharoja.org/no-nos-callamos-mas-y-despues/

Avanza la investigación sobre la acusación a Juanse. (2014b). *El Tribuno*, July 29. Retrieved from www.eltribuno.com/jujuy/nota/2014-7-29-12-54-0-avanza-la-investigacion-sobre-la-acusacion-a-juanse-abuso-sexual-fiscal-fiscalia-victima-juanse

Avigliano, M. (2015). Ni una muerta más. *Página 12*, May 15. Retrieved from www.pagina12.com.ar/diario/suplementos/las12/13-9703-2015-05-15.html

Borrelli Azara, G. (2019). Quereme y vení a bailar. *Página 12*, March 8. Retrieved from www.pagina12.com.ar/179074-quereme-y-veni-a-bailar

Butler, J. (1993). *Bodies that matter: On the discursive limits of sex*. New York: Routledge.

Butler, J. (1999). *Gender trouble: Feminism and the subversion of identity* (2nd ed.). New York: Routledge.

Campaña por el Aborto Legal, Seguro y Gratuito (2019). Retrieved from http://www.abortolegal.com.ar/

Carabajal, M. (2014). Otra víctima del fuego y su machismo. *Página 12*, January 3. Retrieved from www.pagina12.com.ar/diario/sociedad/3-236949-2014-01-03.html

Carrington, K., Hogg, R., Scott, J., Sozzo, M. (Eds.) (2018). *The Palgrave handbook of criminology and the global South*. Abingdon: Routledge.

Coyle, T., & Platt, L. (2019). Feminist politics in the festival space. In J. Mair (Ed.), *The Routledge handbook of festivals*. Abingdon: Routledge.

de Jong, A. (2017). Rethinking activism: Tourism, mobilities and emotion. *Social & Cultural Geography*, *18*(6), 851–868.

de los Santos, G. (2016). Chiara Páez, el crimen de la adolescente que disparó las marchas de Ni una menos. *La Nación*, June 3. Retrieved from www.lanacion.com.ar/seguridad/rufino-chiara-nid1905389

Denuncian a Fidel Nadal por maltrato y acoso. (2018). *Rolling Stone*, July 6. Retrieved from www.lanacion.com.ar/espectaculos/musica/denuncian-a-fidel-nadal-por-maltrato-y-acoso-nid2150843

El derrape de Juanse al defenderse de la acusación de abuso sexual en Jujuy. (2015). *La Voz*, April 9. Retrieved from https://vos.lavoz.com.ar/poprock/el-derrape-de-juanse-al-defenderse-de-la-acusacion-de-abuso-sexual-en-jujuy

El festival cordobés GRL PWR presenta su line-up compuesto íntegramente por mujeres. (2019a). *IndieHoy*, February 18. Retrieved from https://indiehoy.com/noticias/festival-cordobes-grl-pwr-presenta-line-up-compuesto-integramente-mujeres/

El Senado aprobó el proyecto de ley de cupo femenino en festivales de música. (2019b). *Perfil*, May 22. Retrieved from www.perfil.com/noticias/sociedad/el-senado-aprobo-el-proyecto-de-ley-de-cupo-femenino-en-festivales-de-musica.phtml

El Festival GRL PWR con más de 20 bandas en su grilla. (2019c). *Perfil*, March 10. Retrieved from www.perfil.com/noticias/cordoba/el-festival-grl-pwr-con-mas-de-20-bandas-en-su-grilla.phtml

Episodio 37. (2018). *Caja de Herramientas*, June 13. Retrieved from https://audioboom.com/posts/6931080-episodio-37-programa-completo-caja-de-herramientas-marilina-tortillera-alta-paja

Eruca Sativa hace historia en el mítico festival Cosquín Rock 2019. (2019). *Tango Diario*, February 15. Retrieved from https://tangodiario.com.ar/eruca-sativa-hace-historia-en-el-mitico-festival-cosquin-rock-2019/

Evento. (2019). *AlPogo*. Retrieved August 30, 2019, from https://alpogo.com/evento/festival-grl-pwr-898

Festival GRL PWR. (2019). *Facebook*. Retrieved August 30, 2019, from www.facebook.com/festivalgrlpwr/

Festival GRL PWR. *Official Site*. Retrieved August 30, 2019, from http://festivalgrlpwr.com/

Festival GrlPwr 2019: así es la grilla completa del evento 100% femenino. (2019). *La Voz*, February 18. Retrieved from https://vos.lavoz.com.ar/musica/festival-grl-pwr-2019-asi-es-la-grilla-completa-del-evento-100-femenino

Festival GrlPwr: mujeres, músicas y talentosas. (2019). *CBA24*, February 19. Retrieved from www.cba24n.com.ar/mujeres-rockeras-y-muy-talentosas/

Freijo, F. (2019). GRL PWR: el aborto legal y la diversidad fueron las banderas de un festival que reivindicó el talento femenino. *Infobae*, April 19. Retrieved from www.infobae.com/cultura/2019/04/19/grl-pwr-el-aborto-legal-y-la-diversidad-fueron-las-banderas-de-un-festival-que-reivindico-el-talento-femenino/

Gentil, A. (2019). La hora de las femibandas: rock, punk, recitales y feminismo. *Noticias*, May 15. Retrieved from https://noticias.perfil.com/2019/05/15/la-hora-de-las-femibandas-rock-punk-recitales-y-feminismo/

Goldman, E. (1934). *Living my life*. New York: Knopf.

Gordillo, F. (2019). Vuelve el festival Grl Pwr con una grilla compuesta exclusivamente por mujeres. *LM Diario*, February 18. Retrieved from https://lmdiario.com.ar/noticia/131280/vuelve-el-festival-grl-pwr-con-una-grilla-compuesta-exclusivamente-por-mujeres

Hicieron lugar a la probation solicitada por Gustavo Cordera. (2019). *Fiscales.gob.ar*, April 4. Retrieved from www.fiscales.gob.ar/genero/hicieron-lugar-a-la-probation-solicitada-por-gustavo-cordera/

"Hicimos esto para dejar de estar solas": nuevo descargo de las denunciantes de Onda Vaga. (2018a). *La Nación*, December 11. Retrieved from www.lanacion.com.ar/espectaculos/hicimos-esto-dejar-estar-solas-nuevo-descargo-nid2201488

Imputados por abuso sexual. (2014a). *El Tribuno*, July 26. Retrieved from www.eltribuno.com/jujuy/nota/2014-7-26-0-0-0-imputados-por-abuso-sexual-fiscalia-juanse-abuso-sexual-agravado-por-elnumero-de-participantes-ratones-paranoicos-orden-de-captura

Infobae. (2018). *El grupo Onda Vaga rompió el silencio tras las acusaciones de abuso sexual, Nov. 9*. Retrieved from https://www.infobae.com/teleshow/infoshow/2018/11/09/el-grupo-onda-vaga-rompio-el-silencio-tras-las-acusaciones-de-abuso-sexual/

Infoleg. (2019). Retrieved August 30, 2019, from http://servicios.infoleg.gob.ar/infolegInternet/anexos/15000-19999/16546/texact.htm#15

Instituto de Proyección Ciudadana de la Ciudad de Buenos Aires. (2018). *Estadísticas sobre Acoso Callejero en CABA*, November 14.

José Palazzo: "No hay suficientes mujeres con talento a la altura del Cosquín Rock". (2019a). *Perfil*, February 12. Retrieved from www.perfil.com/noticias/sociedad/jose-palazzo-dijo-no-hay-suficientes-mujeres-con-talento-altura-del-cosquin-rock.phtml

José Palazzo y la polémica por el cupo femenino en el Cosquín Rock: "Hoy la escena no tiene tantas bandas de mujeres". (2019). *TN*, February 12. Retrieved from https://tn.com.ar/show/basicas/jose-palazzo-y-el-cupo-femenino-en-el-cosquin-rock-hoy-la-escena-no-tiene-tantas-bandas-de-mujeres_940004

Juanse, exlíder de los Ratones Paranoicos, imputado por abuso sexual agravado. (2014). *La Voz*, June 26. Retrieved from www.lavoz.com.ar/sucesos/juanse-exlider-de-los-ratones-paranoicos-imputado-por-abuso-sexual-agravado

Juanse pidió disculpas a Jujuy y a sus mujeres. (2015). *El Tribuno*, August 30. Retrieved from www.eltribuno.com/jujuy/nota/2015-8-30-0-0-0-juanse-pidio-disculpas-a-jujuy-y-a-sus-mujeres-inadi-ante-la-denuncia-de-liliana-fellner-juanse

Lahiteau, L. (2018). A la violencia machista en el rock argentino le llegó su #Time'sUp. *Revista Arcadia*, July 6. Retrieved from www.revistaarcadia.com/musica/articulo/denuncias-de-violencia-machista-rock-argentino-niunamenos-timesup-metoo/69914

Lamond, I., & Spracklen, K. (2014). *Protests as events: Politics, activism and leisure*. London: Pickering and Chatto Publishers.

Llega el Festival GRL Power con una grilla compuesta exclusivamente por mujeres. (2019). *Filo News*, February 18. Retrieved from www.filo.news/musica/Llega-el-Festival-GRL-Power-con-una-grilla-compuesta-exclusivamente-por-mujeres-20190218-0050.html

Los Espíritus: Maxi Prietto publicó su descargo de las acusaciones de abuso. (2019). *Tiempo Argentino*, March 21. Retrieved from www.tiempoar.com.ar/nota/los-espiritus-maxi-prietto-publico-su-descargo-de-las-acusaciones-de-abuso

Marilina Tortillera. (2019a). *Bailame*. Retrieved August 30, 2019, from www.bailame.com.ar/marilina-tortillera.html

Marilina Tortillera. (2019b). *Facebook*. Retrieved August 30, 2019, from www.facebook.com/FiestaMarilinaTortillera/

Massey, D. (2013). *Space, place and gender*. Oxford: John Wiley & Sons.

McGillivray, D., & Jones, J. (2013). Events and resistance. In R. Finkel, D. McGillvray, G. McPherson, & P. Robinson (Eds.), *Research themes for events* (pp. 129–141). London: CABI.

Ni Una Menos. (2017). Retrieved from http://niunamenos.org.ar/quienes-somos/carta-organica/

Palazzo: "Pido disculpas si alguna mujer se sintió ofendida por el malentendido". (2019). *Télam*, February 14. Retrieved from www.telam.com.ar/notas/201902/332868-jose-palazzo-productor-cosquin-rock-declaraciones-artistas-mujeres.html

Por una denuncia de abuso, el cantante Maxi Prietto fue desvinculado de Los Espíritus. (2019). *Clarín*, March 2. Retrieved from www.clarin.com/espectaculos/denuncia-abuso-cantante-maxi-prietto-desvinculado-espiritus_0_wIoHOp-_P.html

Procesaron al cantante de La Ola Que Quería Ser Chau. (2017). *Rolling Stone*, April 26. Retrieved from www.lanacion.com.ar/espectaculos/procesaron-al-cantante-de-la-ola-que-queria-ser-chau-nid2017555

Rojek, C. (2005). *Leisure theory: Principles and practice*. London: Palgrave Macmillan.

Rose, G. (1993). *Feminism and geography: The limits of geographical knowledge*. Minneapolis: University of Minnesota Press.

Scraton, S., & Watson, B. (1998). Gendered cities: Women and public leisure space in the "postmodern city". *Leisure Studies*, *17*(2), 123–137.

Se sinceró el cantante de Salta la Banca: fuerte carta abierta en medio de un show. (2017). *Clarín*, October 31. Retrieved from www.clarin.com/espectaculos/fama/sincero-can tante-salta-banca-fuerte-carta-abierta-medio-show_0_ryndhArC-.html

Sharpe, E. (2008). Festivals and social change: Intersections of pleasure and politics at a community music festival. *Leisure Sciences*, *30*(3), 217–234.

Solicitaron enjuiciar al músico Gustavo Cordera por incitación a la violencia contra las mujeres. (2018). *Fiscales.gob.ar*, June 7. Retrieved from www.fiscales.gob.ar/genero/solicitaron-enjuiciar-al-musico-gustavo-cordera-por-incitacion-a-la-violencia-contra-las-mujeres/

SOY. Agenda. Del viernes 16 al jueves 22 de agosto. (2019). *Página 12*, August 16. Retrieved from www.pagina12.com.ar/212009-agenda

Timerman, J. (2018). El debate inesperado sobre el aborto en Argentina. *New York Times*, May 19. Retrieved from www.nytimes.com/es/2018/05/19/opinion-timerman-misogi nia-feminicidio-debate-aborto-argentina/

Tras las denuncias de abuso, Los Espíritus se retractaron y volvieron a incluir a Maxi Prietto en la banda. (2019). *Filo News*, March 17. Retrieved from www.filo.news/actualidad/Los-Espiritus-se-retractaron-y-volvieron-a-incluir-a-Maxi-Prietto-en-la-banda-20190317-0008.html

Unas vagas disculpas. (2018). *Página 12*, November 10. Retrieved from www.pagina12.com.ar/154576-unas-vagas-disculpas

Vegh Weis, V. (2018). *Marxism and criminology: A history of criminal selectivity*. Chicago: Haymarket Books.

Vegh Weis, V. (forthcoming). It is time! Towards a southern convict criminology. In J. I. Ross (Ed.), *The future of convict criminology*. London: Routledge.

Walas, tras su show en Mendoza: "Dije una barbaridad y tengo que pedir disculpas". (2016). *Silencio*, April 17. Retrieved from https://silencio.com.ar/noticias/lo-ultimo/walas-pide-disculpas-dichos-mendoza-7185/

Ya no nos callamos más. (2019). Retrieved from https://yanonoscallamosmas.wordpress.com/

11 The polite cowboy or the wild, wild west

Strategic approaches to reducing gender-based rodeo violence through grassroots civic mobilisation

Mylynn Felt and Maria Bakardjieva

Introduction

The Calgary Stampede (CS) is "an annual ten-day festival built around a world-class rodeo, a modern midway, and a frontier western theme that spills beyond the [designated] grounds to the city itself" (Foran, 2008, 2). In the minds of the citizens of Calgary, the Stampede represents "ten mad days in July" (Foran, 2008, 2), days that different people respond to in widely divergent ways – from a desire to stay as far away as possible, to passionately embracing the clothing, the food and the activities, as well as the cultural symbolism of the festival. The CS is a year-round, high-power operation instituted on a non-profit principle, including local dignitaries on its board of directors and having close links to the city's municipal administration.

The origins of the event lie in an early agricultural fair, the Calgary Exhibition, dating from 1886. The Stampede is believed to have high commercial value; hence, the City of Calgary, the Chamber of Commerce, local businesses, livestock associations, tourist agencies and local media all actively support and propagate the event. The popularity of CS has grown over the years. In 2017, it was visited by 1.2 million people (CBC, 2017). In its current form, it represents a trade show, a display of cultural mythology and a carnival all in one. The carnival spirit, in particular, sweeps over the whole city and involves suspension of the constraints of everyday life, relaxation of social norms, role reversals, 'goofy' behaviour and morally questionable activity, such as affairs. The slogan "it's not cheating; it's stampeding' is one often mocked but also associated with the event. Participants typically don western garb, pointy boots, cowboy hats and wide-open necklines. The celebration features parades, pancake breakfasts, powdered mini doughnuts, musical performances, carnival rides and fireworks. Initially envisioned as a celebration of the western Canadian frontier experience, the spirit of the Stampede was gradually overtaken by Wild West symbolism stemming from Hollywood movies, and since the 1960s, it has focussed primarily on "the generic western myth" (Foran, 2009).

The loosening of social norms is taken by some as a licence to see many of the women in the Stampede crowd who are there to enjoy the festivities, or there to do their jobs at bars, as "prostitutes and bar girls" like those depicted in Hollywood westerns. In an article detailing the sexual harassment women face during Stampede, Lyndsie Bourgon (2015) noted, "For the duration of Stampede, the city's already-simmering frat culture is on display, given the permission it needs to shout loudly". The notion of frat culture refers to the hypermasculinity associated with university fraternities. The article cited a research project studying women, sex work and the Stampede conducted by Calgary academic Kimberly Williams, who issued a stronger verdict: "I'm suggesting that there is a culture in the Stampede that makes violence against women acceptable and normal" (Bourgon, 2015). The association of these practices with a 'legendary' public event sponsored by civil and corporate authorities sends the wrong message that can affect gender relations far beyond the Stampede itself.

According to Bourgon's article, some Calgary community service organisations (CSOs) working in the area of preventing violence against women admitted that they would have liked to organise a campaign to confront the issue of Stampede-time sexism and abuse, but they were worried about "annoying those that make up the Stampede board and corporate Calgary because those are also the supporters of the agencies" (Bourgon, 2015). Organising such a campaign is exactly the role that the grassroots initiative examined in this chapter took upon itself. Concern about the sexism and sexual harassment associated with the Stampede led local women to design an initiative to create a safer environment. Its key instrument was the Twitter hashtag #SafeStampede. Campaign organisers called on Stampede participants to elevate their behaviour. The organisers included women who had found each other through private Facebook accounts where they identified people sharing similar concerns. They were ordinary citizens, not employees or activists of any organisations. In what follows, we refer to them as grassroots organisers. Much of the original planning of the campaign occurred online through what the organisers saw as 'safe spaces' where strategising without public criticism is possible.

The campaign that we focus on occurred in Calgary, Canada, between 2015 and 2017. It was initiated by women who wanted change in cultural norms concerning gender relations. It heavily relied on Twitter and drew together a dedicated group of ordinary citizens who turned into activists in the process. Further, it forced public bodies to reconsider cultural aspects that they had previously taken for granted with a special reference to an iconic public event held annually, the biggest rodeo in the world, the CS.

Our goal is to track the ways in which grassroots organisers employed digital media to call for a cultural change with respect to the treatment of women in the context of the CS. Our study follows the evolution of the links between CSOs and grassroots initiators and seeks to understand how the interaction between them affected the mediation of the initiative. We ask, How did grassroots campaign initiators and established CSOs cooperate and divide labour in their efforts to reach wide audiences and transform cultural attitudes, norms and behaviour? How did their respective involvement influence the mediation of the campaign? What were the costs and benefits of the partnership between these actors with respect to the persistence and reach of their message? To conceptualise this side of the analysis, we borrow ideas from the sociological literature on social movements and particularly the notion of 'institutionalisation', which we apply with caution given the small-scale nature of the movement under examination.

Genesis of the campaign

Organisers who started the #SafeStampede initiative first came together over a related issue in April 2015. They were successful in gaining broad awareness and support for curbing sexual harassment during the National Hockey League playoffs using #SafeRedMile. With the help of news media attention and a couple of key instances demonstrating the problem, just one day after launching the campaign, Calgary Flames team management held a press conference calling on fans to stop harassing women. During game breaks, they also made public service announcements about using respectful behaviour when donning the jersey or cheering for their team. The organisers were thrilled by this rapid response and continued their efforts throughout the rest of the playoffs. By May of 2015, supporters were responding on Twitter to #SafeRedMile with "hope [that] you can continue the conversation through Stampede".

Six of the #SafeRedMile organisers formed a group that met face-to-face and worked to plan the #SafeStampede initiative just weeks before the event began in 2015. During the first planning meeting, Emma, an organiser with a legal background and corporate connections, argued that they should not vilify CS itself but should try to gain the Stampede board as an ally. They also wanted to do more than just identify a problem. They wanted to find ways that the public could address the issue. For this purpose, Kenna suggested they partner with a local CSO, the Calgary Sexual Health (CSH) Centre so that awareness could grow into education by tapping into the organisation's resources. Pam, the director of CSH, was happy to help but was also eager to keep the grassroots organisers central to planning and action. CSH created a website that defined the problem and promoted resources, such as bystander intervention training and WiseGuyz, a programme for junior high school boys that teaches human rights, sexual health, gender and positive relationships (Spence, 2018). During planning meetings, Elizabeth agreed to assume the role of Twitter spokesperson for the campaign. Both the grassroots organisers and CSH reached out to the general public through Twitter, Tumblr and mainstream news media, all pointing back to the website professionally maintained by CSH.

In the following analysis, we will trace the social media activities of the Safe Stampede campaigners as the steering of the initiative moved from a grassroots group into the hands of formal CSOs. We set out to find how progressive institutionalisation of the initially spontaneous grassroots movement affects the mediation of its message and public reach.

Trajectory of collective action campaigns

Collective action campaigns designed to address broad social problems require ongoing efforts to effect social change. Resource mobilisation and political process theories argue that collective action evolves along four stages: 1) diffusion, 2) exhaustion, 3) either radicalisation or institutionalisation and 4) restabilisation (Tarrow, 2011). Exhaustion is likely for initiatives that arise from the grassroots because political claims making and mobilisation require significant investment of time, effort and resources by individuals who must also meet the daily demands of life. In order for collective action to persist over longer periods, it needs to move from loose collectives to more formalised organisations.

For Kriesi (1996), this evolution is more complicated. Social movement organisations (SMOs) tend to follow one of four trajectories in the third stage: 1) institutionalisation, 2) commercialisation, 3) involution or 4) radicalisation (Kriesi, 1996). In the process of institutionalisation, SMOs see more formalised internal structure, as well as moderation of goals. Becoming more like a party or interest group, the process stabilises resource flow, conventionalises the action repertoire and creates communities of common interest. Commercialisation of a SMO describes the transformation into a profit-making enterprise or service organisation. Involution leads to exclusive emphasis on social incentives, while radicalisation reflects an escalation in goals and tactics along with reinvigorated mobilisation.

To these potential outcomes, this case presents a fifth trajectory, one we call formalisation, as pictured in Figure 11.1. In this process, grassroots organisers partner with formal organisations, such as CSOs, incrementally until the initiative is managed entirely by formal organisations that exist independently of the campaign. Unlike institutionalisation, in which an initially spontaneous

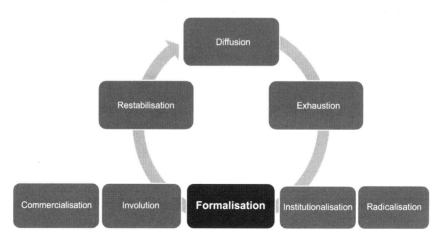

Figure 11.1 Processes of contention

and loosely coordinated grassroots movement consolidates its internal structure, formalisation sees responsibility for a grassroots campaign turned over to existing formal organisations.

A movement's communication processes look different depending on the stage. To examine the changes occurring in these processes, we use Cammaerts' (2018) conceptual model, which "encompasses the various moments in which media and communication are implicated in protest and social change" (p. 7). Labelled 'the circuit of protest', Cammaerts's model comprises four key moments: production of movement claims and frameworks, self-mediation practices, representations of the movement by mainstream media and reception of the messages produced by the movement and the broader public (Cammaerts, 2018, 7). This approach values not only processes of production but also those of media consumption and the reception of meaning, each of which are critical aspects of civic activism.

In a case study examining a transnational feminist campaign called "Take Back the Tech!", Pavan (2017) identifies three primary stages as grassroots organisers progress along the path of institutionalisation: 1) mild institutionalisation, 2) inclusion and 3) consolidation. In mild institutionalisation, organisers begin working with both institutional and non-institutional actors. In the stage of inclusion, the concentration shifts more to institutional actors. The final stage of consolidation occurs when a campaign is run nearly entirely by institutional actors. From 2015 to 2017, a similar pattern emerged as grassroots organisers of Safe Stampede increasingly relied on and deferred to formal organisations. The peculiarity in our case was that those organisations did not emerge from the movement itself but already existed and willingly joined the grassroots activists in the pursuit of their goal.

Methodology

This case focusses on #SafeStampede, 2015–2017. The research model combines both qualitative methods with online data analytics to reveal the macro and the micro. The analytical framework for this analysis is informed by social movement theory on processes of institutionalisation, as well as Cammaerts's circuit of protest model (2018). The circuit of protest focusses on the message production, self-mediation practices, mainstream media representation and public reception and response. We interviewed eight key organisers between September 2015 and April 2017, asking questions concerning the processes of message generation, sharing and circulation. Because the organisers' primary means for reaching out to the public was through Twitter, we collected tweets utilising the Twitter Capture and Analysis Toolset (Borra & Rieder, 2014) with search terms 'safe' and 'stampede' as well as '#safestampede'. Metadata[1] and textual analysis were conducted through spreadsheets. We examined all links present in individual tweets; they connected to websites; other social media platforms, such as Instagram; and mass media articles. Further searches for news articles were conducted through the Factiva database. This search was limited to Canadian news publications utilising the search terms 'safe' and 'stampede' and 'Calgary' from June 15, 2015, to July 31, 2017.

Findings

Table 11.1 conveys Twitter data evidence of the organisational progression of this initiative at different stages.

Mild institutionalisation: 2015 Safe Stampede campaign

The first year consisted primarily of grassroots organisers partnering with CSH. This trajectory was deliberate. The organisers reached out to CSH in their earliest planning. In 2015, one of the organisers, Elizabeth, said,

> We wanted to keep it fairly grassroots, but Calgary Sexual Health thought they could back us up a little bit with resources and then hopefully it would snowball into something larger the second year – that they could take on a more formal capacity.

Another organizer, Kenna, reflected that they did not have the time necessary to pull together professional promotional materials in the first year, but they knew that part of the long game would include more partners for the second or third year and beyond. Each year of this campaign progressed according to Pavan's (2017) stages of mild institutionalisation, inclusion and consolidation.

Inclusion: 2016 Safe Stampede campaign

In the second year, fewer grassroots organisers participated, and more formal organisations such as NextGenMen and the Society for the Advocacy of Safer Spaces supported the campaign. Elizabeth, the grassroots organizer who acted as a spokesperson for the campaign in the first year, indicated a desire for CSH to take on a more formal role after the first year, citing exhaustion, work pressures and her own personal reputation to consider. Elizabeth recognised that partnering with CSH not only lent legitimisation to the online campaign

Table 11.1 Tweet summary

	Mild institutionalisation	*Inclusion*	*Consolidation*
Year	2015	2016	2017
(N) Tweets	1,152	503	661
(N) Users[2]	665	327	318
Types of Media Links	News 11	News 10	News 8
	Advocacy resources 1	Advocacy resources 1	Advocacy resources 8

but also provided the resources and training to extend their efforts beyond online claims and broaden their action repertoire. Pam, executive director of CSH, described the 2016 campaign this way:

> I think this year, for me, the focus felt a bit more institutional, from the perspective that the Calgary Stampede was actually involved in the news conference. They did tattoos, so they actually put a little bit of skin in the game. Then we had the Minister of the Status of Women be a part of our news conference. ... I don't feel like it lost its grassroots feel to it, but it actually felt like it actually had taken that next step: institutions being interested.

Consolidation: 2017 Safe Stampede campaign

By the third year, the campaign was run nearly entirely by CSH in partnership with CS and the government. This stage of 'consolidation' included community partners, such as the Calgary Domestic Violence Collective, Calgary Women's Shelter and the YWCA. After the 2016 campaign, Pam, the CHS director, indicated that she fully intended to bring back the women who initiated #SafeStampede, yet none participated in 2017. Though some of the original organisers retweeted and commented on the campaign, they did not participate in press releases or appear prominently in social media. Although online discourse on Safe Stampede claims was greatly reduced by this stage, message framing was superseded by a broadened repertoire of action, such as trainings and temporary tattoos as endorsements of the campaign.

Twitter engagement

As Table 11.1 shows, the first year of #SafeStampede engaged the largest audience on Twitter, with 665 accounts participating and 1,152 individual tweets. This year also saw the greatest number of linked news articles. Some of the links pointed to what we call 'advocacy resources', which are websites and brochures designed by CSOs to help members of the public. The only advocacy resource link in the first two years of the campaign directed readers to the website hosted by CSH and designed in cooperation with the original grassroots organisers. The website served as a landing page where advocates could more thoroughly frame the message. Each year saw fewer Twitter users participating in the campaign, less than half between the first and third years. By the stage of consolidation, the number of news articles that tweets linked to had also fallen to just eight. At the same time, the number of advocacy resources had grown to include Calgary Women's Shelter handbook "Respecting and Listening to Victims of Violence" and blogs from the Calgary Domestic Violence Collective and YWCA Calgary. This shows a substantive drop in public engagement and a higher reliance on the ready-made content published by formal organisations.

#SafeStampede circuit of protest

The #SafeStampede initiative was more than just a Twitter campaign. The following results are examined through the circuit of protest (Cammaerts, 2018) in a chronological manner through the stages of institutionalisation.

Production

The first stage of the circuit of protest is production. It is at this level that

> social movement actors produce or encode meaning through discourses and frames, whereby the former represents inherent contingency and, the latter, strategic attempts to fix meaning, to establish ideological boundaries and to construct a "we" that is juxtaposed to a "them".
>
> (Cammaerts, 2018, 33)

Production, in 2015, was focussed on defining the problem. Many tweets from organisers linked to the article titled, "Herd Mentality" (Bourgon, 2015), which laid out the problem and justified the need for specifically grassroots mobilisation on the issue. The grassroots organisers made a strategic decision to approach Stampede board members and assure them that they did not intend to hurt CS but rather hoped to redirect the Stampede image to a family-friendly event and away from the hyper-sexualised representations they argued others had used to hijack the brand. Organisers called on the public to share their own experiences of harassment, sexism or assault on an anonymous Tumblr page. They argued that it was time to change the culture to one of consent and respect. Kenna said the first year was really about starting the ball rolling and having a conversation about what is neighbourly behaviour. Pam characterised the message of 2015 to be that there should not be an exemption for ten days when people feel free to exhibit bad behaviour.

From the beginning, Elizabeth reflected that partnering with organisations caused the grassroots organisers to moderate their claims: "When you are working with other people you do have to police yourself a little bit and police your tone and realise there are other people involved and you can't necessarily speak for everybody". Organisers met with CSH in planning sessions. Elizabeth noted,

> It's important to have your message and stick to it, or at least [know] what your message is and know what your goal is ... do not deviate from it too much, but I think you also don't want to get into a talking points situation where you're not listening to people, because that's the point of grassroots and being social media. It's a dialogue with people; it's not a PR campaign.

She recognised that formal partners had a greater capacity to run a social media campaign but valued grassroots organising and online discourse.

In the second year, tweets by grassroots organisers argued that they were looking to battle the normalisation of sexual harassment at Stampede. Elizabeth described the difference between claims of the first and second year by saying that in 2015 was about alerting people that there was a problem, and 2016 was about instituting some solutions. One grassroots organizer, Gina,[3] was pleased that CSH was "doing something more. I feel like they've really taken that ball of the bystander training and gone with it and seem to have done great work in that area".

In the third year, grassroots organisers did not make claims mediated through social media or news media on behalf of the campaign. CSH and NextGenMen worked to spearhead the initiative. CSH shared the work they had been doing to train Stampede staff on bystander intervention, and they announced government funding that would make it possible to continue such efforts throughout the year. Jake, from NextGenMen, saw #SafeStampede on Twitter and had an idea for an interactive experience that could enrich the campaign. NextGenMen produced a series of tweets quoting participants in the "This Is What It Feels Like" (TIWIFL) exhibit, which involved inviting about 600 pedestrians to spend a few minutes in a trailer experiencing recordings of street harassment while looking at a mirror. Participants were then allowed to write and make comments on their reactions to the experience.

Self-mediation

Self-mediation refers to the "variety of mediation practices using textual, audio and visual formats, distributed offline and online, locally, nationally and even transnationally"; it is focussed not as much on the symbolic as on the material aspects of organising (Cammaerts, 2018, 33). The first year of organising emphasised messages mediated through Twitter, Tumblr and the website organisers created with CSH. The grassroots organisers found that the Tumblr page did not gain the traction they were hoping for. Kenna reflected that hundreds of women responded on Facebook, but when they were invited to make anonymous Tumblr statements describing the harassment they had received, few were willing to move to a different platform. While Tumblr fostered more desirable affordances for sharing, the routine media practices of supporters shaped the mediation decisions of organisers.

The second year also focussed on the website and Twitter, but all use of the Tumblr page died off. Spearheaded by CSH, organizers hosted a media launch for #SafeStampede, inviting reporters to hear from a Stampede representative, the minister on the status of women, CSH representatives and one of the grassroots organizers. The issue of violence against women was moving centre stage for the organizers of CS.

A similar media launch was hosted in 2017 but this time without any of the grassroots organisers. Though the third year also included links and mentions of the website, more emphasis was placed on bridging to CSH's new initiative called #CalgaryGetsConsent. NextGenMen was more heavily involved in 2017, as they hosted the TIWIFL project that had a central face-to-face component, as well as used scheduled tweets and Instagram posts sharing its results. This content accounted for most of the #SafeStampede hashtag use. Partnering organisations also linked to their own resources and blogs, using Twitter to share opinions on sexual harassment.

Representation

Representation focusses on how mainstream media actors and journalists, situated outside the movement, cover and convey the messages and imagery framed by movement organisers (Cammaerts, 2018). In this regard, news articles in the first year of this campaign aided in establishing value for the initiative. One article asked, "Does Calgary Need #SafeStampede for a Harassment-Free Week?" One reporter tweeted the rhetorical question, "In case you wondered if we need SafeStampede" while sharing a public challenge on Reddit that called on Stampede revellers to post observed crude behaviour. Articles quoted Pam from CSH, as well as grassroots organisers alike. They promoted the Tumblr page, the website and the hashtag. Reporters noted Twitter endorsements by the mayor and Stampede executive staff. News coverage in the first year was both local and national and quoted more than one of the grassroots organisers. Elizabeth confessed that 100% of the goal in online claims making was to also get mass media attention. Pam expressed concern that reporters did not make the effort to track down the grassroots organisers of the campaign. Instead, she, as an organisational leader, was interviewed ten times that year. She said she repeatedly identified clear spokespeople from the grassroots side of organising because that was what gave the campaign its energy and authenticity, but reporters did not respond.

Near the end of the 2015 campaign, a controversial video showing an explicit sexual act in the parking lot of Stampede grounds gained broad attention through social media. Reporters used that as evidence to emphasise the merit of #SafeStampede. One radio host interviewed the woman who was one of the subjects in the video. Though this interview did not discuss #SafeStampede, many tweets connected it with the campaign. Pam argued that in the absence of the Safe Stampede campaign, this still would have been a news story, but with their frame around it, they were able to direct the conversation to one about consent and slut shaming.

In 2016, most early news coverage developed from the media launch hosted by CSH. Reporters claimed that three partners, the province, the Stampede board and CSH, spearheaded the project. Though one of the grassroots organisers was present for the media launch, she was not quoted or mentioned in most of the initial articles. She was only quoted as someone supporting the concerns over sexualised advertising raised by members of the public. Minister for the Status of Women Stephanie McLean appeared in articles related to the Stampede, discussing the importance of active bystanders. Pam argued that the involvement of official figures led to the proliferation of the campaign's message across mainstream media, something her organisation could never afford through advertising.

In 2017, the news focus was on the new funding CSH received to launch its year-long campaign #CalgaryGetsConsent. These articles also shared numbers on bystander intervention training that CSH conducted with Stampede venue employees. The dominant article that year only mentioned #SafeStampede in the final line. Representatives quoted include the government minister, the head of CSH and a Stampede representative. Unlike previous years, no controversies or concerns raised by members of the public initiated further reporting beyond the initial media launch.

Reception

Public opinion is "another potential influencer of mainstream media representations and political actors in a democracy" (Cammaerts, 2018, 34). Reception in the social media environment does not only indicate how public opinion changes as a result of civic initiatives but also comprises the way that non-activist citizens react to social movement claims.

The first year of mild institutionalisation (2015) saw the most debate by the public on this campaign. In addition to the anonymous accounts of past victimisation shared on the Tumblr account, Stampede attendees took to Twitter to witness "lots of unnecessary harassment happening tonight – remember #safestampede and consent". Others claimed that "just because I dress up in shorts, cowboy boots and hat doesn't mean I'm asking for it". Early in the Twitter discourse this year, members of the public challenged the boundaries of the initiative by questioning if it only applied to women or if the campaign also sought to protect men from sexual harassment. Others questioned how it could be a Safe Stampede if the animals, particularly the horses involved in chuck wagon racing, were at risk. Elizabeth viewed these challenges as a derailment tactic. "People don't want the issue to be about women because then they have to admit that women are being harassed on the street during Stampede and don't feel safe and nobody wants to admit that". Kenna compared the efforts to stretch #SafeStampede beyond more than just safety for women to detractors of the Black Lives Matter campaign who argue "all lives matter" in retort. Pam shared that someone criticised their efforts as being merely 'slacktivists'. She related that this is one of her greatest fears for online campaigns, and it is a major motivation behind bystander intervention training, because while social media consciousness raising is important, she was concerned it just was not enough.

CS posted tweets calling for participants to demonstrate western hospitality, to show respect and "western values". Of course, the notion of 'western values' is not explained in these tweets and could be interpreted as the source of the problem under scrutiny as notions of cowboy culture and the wild, wild west grant licence for sexist conduct. When the controversial video showing sexual acts performed at Stampede grounds gained broad attention on the website Reddit, #SafeStampede organisers called on the r/Calgary[4] moderators to act as civic leaders and remove the video to protect the people involved. The short video, made available on Reddit and YouTube, was viewed over a million times in just a few days. Once the woman in the video was identified, she experienced overwhelming slut shaming and harassment. She responded to this with her own YouTube channel and radio interview to defend her actions. The men involved were never publicly named or harassed for their part in the incident. The Twitter conversation between Emma and the Reddit moderator drew in many members of the public who took sides on whether the video belonged in the public gaze and whether there was a duty to protect those involved.

Elizabeth was the sole #SafeStampede champion who took it upon herself to intervene in that controversy on behalf of the initiative. She, however, felt frustrated by the limitations of Twitter as a platform for handling controversial situations like this. "I felt terrible for her [the woman in the video], and I think she didn't know how to react. Explaining that to people who don't feel empathy for her in 140 characters is not possible". Elizabeth described that experience as one that drained a lot of energy and left her emotionally battered. It made her reassess and scale down her involvement in the aftermath. In our interviews conducted a year after these events, other organisers shared that they were surprised by their own unwillingness to speak out in Elizabeth's defence and decisively support her position in public. They cited fear of also being drawn into the discord. Kenna said she was really worried about Elizabeth and the girl who was at the centre of the video. At the same time, she felt the fact that "Safe Stampede and that conversation was already in place meant that there was a platform for us to be able to address that". Pam admitted that CSH would never engage in a live controversial online discourse to the extent Elizabeth did. They assume more of a traditional spokesperson role by taking questions from reporters.

At the stage of inclusion (2016), several formal partners supported #SafeStampede, and the public sometimes responded more to the support than to the campaign. When Calgary's mayor promoted the hashtag, many responded to his tweet with disappointment that the 'safety' of the campaign did not extend to the animals as well. The CS Twitter account announced that "Calgary's reputation as a warm, safe and welcoming city should be at its most powerful during #stampede2016". The response from the members of the community was multi-voiced and not necessarily agreeable. Some tweeted that they "avoid Stampede every year" and that Stampede needs to maintain safer spaces. One person opened a Twitter account for the sole purpose of posting sarcastic claims under the #SafeStampede hashtag in order to challenge the initiative. Another person saw the campaign as an opportunity to promote her self-defence classes. Members of the public tweeted pictures donning the temporary tattoos handed out by one of the large Stampede venues, Nashville North, that participated in CSH's bystander intervention training.

Safe Stampede supporters called out three specific instances of sexualised advertising. The one that received the most attention was a set of three posters conveying sexualised cowboy culture on the wall of a popular restaurant. One Instagram post tagged the brands associated with the ad as well as several news organizations. This triggered another wave of abusive comments targeting Elizabeth, who gave a televised interview regarding the advertisement. This is where she saw the value of formal partners.

> If I'm out there all by myself screaming about sexual harassment, I look like the outsider. I look like the radical. ... If I'm talking about sexual harassment and my pal, the mayor, is behind me and the guy from the Stampede is behind me, and the Minister for the Status of Women is behind me, companies, restaurants I eat at, things people think are cool, then the one who looks like the outsider and the radical is the one screaming misogynist slurs.

Following the televised interview sharing concerns with sexualised advertising, several members of the public and grassroots organisers reached out to the restaurant and the posters were removed as a result. Pam saw this instance as evidence that "consciousness raising ultimately does have an impact – will ultimately have an impact on how that [Stampede] week is treated".

In the stage of consolidation (2017), the public was less involved in online discourse for #SafeStampede. Hundreds of people experienced the TIWIFL exhibit hosted by NextGenMen, generating what Jake called several uncomfortable conversations, and many public quotes reflecting on the experience were shared by their Twitter account. Diverse individuals of all adult ages participated in the exhibit. However, no 2017 tweets identified specific problems with Stampede as in previous years. CS posted another call for respectful behaviour when individuals don a cowboy hat. These official tweets did not see the same degree of retweets from the public as in previous years.

Discussion

The original grassroots organisers of #SafeStampede arranged the campaign in response to concerns they saw impacting their community during a highly popular annual festival that was supposed to celebrate local culture but was concomitantly reinforcing sexual stereotypes and sexual harassment. The core grassroots organisers had no illusions. One of them, Elizabeth, acknowledged the long process that is required to change cultural norms.

> Obviously, I'd like to wake up and next year, at the first week in July, and have the Stampede be a totally safe event for everybody, you know, that isn't based on a culture of over-sexualization, but that's not going to happen. This is going to take decades to do a full cultural shift. I think we are on the right track. I think it's gone from basically nobody publicly talking about it to a lot of people publicly talking about it.

From the first discussions initialising #SafeStampede, both CSH and the everyday citizens who launched the campaign struggled to find the balance between a formal organisational entity on the one hand and the civic grassroots on the other.

They knew that the change they sought could not happen in a single year and that long-term perseverance would be needed to achieve anything.

In partnering with formal organizations, grassroots organisers did not want #SafeStampede to become a public relations campaign. These concerns reveal an important distinction in the ethos of claims making by grassroots collectives as opposed to institutions. Despite all efforts to maintain the significant involvement of the grassroots instigators, most burnt out after the first year, and by the third year, the campaign moved forward only through formal organisations. Nevertheless, the efforts of organisers managed to effectively place the issue of sexual harassment and assault on the agenda of political and administrative bodies. There were several ways in which the grassroots-formal organisation partnership contributed to the relative success of the initiative. First, spontaneous grassroots upheaval created an opening for formal CSOs to highlight an important social problem that they otherwise shied away from in fear of spoiling their relationship with sponsors. The coordination of activities with relevant CSOs that grassroots initiators sought from the early stages of their campaign gave their micro-movement legitimacy, a solid base of resources and a visible foothold in media space through a professional website and a publicly recognisable spokesperson. CS was confronted by a union of well-known and trustworthy CSOs and the energy and media productivity of impassioned citizens that together represented a convincing claims-making force. Against that background, the problem identified by the campaigners and the rectifying measures they proposed made an impression on the festival's management and legitimised policy change and the investment of attention and resources. The partnership demonstrated that although defending feminist messages in online environments may exact high affective cost and is hard to individually sustain over time, with the support of formal organizations, campaigns for social change can last well beyond the stamina of grassroots organisers. Finally, the later iterations of the campaign that were driven mainly by formal organizations demonstrated a significant decline in online public interest and in the mass media. When everyday citizens led the claim production and mediation process, other citizens responded in a variety of messy and often controversial ways. This contributed to the vigour and visibility of the discussion, the latter enhanced by mass media. In the hands of formal organizations, the online stream of discourse dried up, and the action repertoire employed by campaigners was reduced to formal appeals and important, but narrowly localised, activities, such as bystander training workshops.

The progressive formalisation of the initiative over the years impacted each moment of the circuit of protest. At the stage of claim production, it was evident that the lack of formal organisational affiliation allowed grassroots organisers to speak in highly expressive, emotionally charged authentic messages decrying the sexual abuse occurring during CS. No formal organization working in the area of gender-based violence had been able to resort to the strong language and combative stance that individual citizens generated. This informal production opened the floor for many diverse opinions to be voiced, as well as for conflict to take place.

At higher stages of formalisation, when established organisations took over, message content fell into the more traditional format of public relations. Organisational and corporate employees were the key content producers. The engagement of grassroots citizens in content production was minimal, mostly in the form of re-tweeting or liking. Self-mediation at the later formalised stages employed press conferences, mass media interviews and strategic use of social media platforms by organisations. Reception was active, with free range for oppositional readings and reactions stirring controversies at the grassroots stage and more reserved and passive with advancing formalisation. Clearly, from a mediation perspective, there was an ideal balance in the earlier stage of formalisation where grassroots energy and organisational resources and established legitimacy worked together to create a strong momentum for both broader public participation and attracting the attention of the institutional players.

Conclusion

The #SafeStampede initiative sought to shift cultural norms of hyper-sexualisation and harassment associated with an annual western rodeo. Campaigns designed to renew on a regular basis are prone to increasing exhaustion on the part of organisers. This is particularly the case with grassroots organisers who must also negotiate the demands of careers, family and everyday life. The organisers of this campaign successfully managed to place this issue on the agenda of news agencies, government bodies, the board of directors for the event itself and venues where community celebrations spill past Stampede grounds. Importantly, these efforts proved successful due to the partnerships between grassroots organisers and CSOs dedicated to addressing gender-based violence. By partnering with CSH, the grassroots organisers were able to do more for the public than just raise short-term awareness about a problem. Many of the venues tied to Stampede oriented their attention towards the issue and enhanced their ability to prevent harassment and assault through bystander intervention training. Furthermore, CSOs, such as NextGenMen and local women's shelters, were able to offer various platforms for a regular public conversation. The process of formalisation in this case brought many benefits. By partnering with formal organisations and strategically working with festival management, the grassroots organisers demonstrated tactical decisions that improved the reach, legitimacy and longevity of their campaign.

There were also, however, costs for the loss of grassroots organisation. In the first and second years of this campaign, awareness raising led to greater online public engagement. When supporters called out sexualised advertising, public pressure led to the removal of ads. In the third year, when grassroots organisers were minimally involved, the nature of Twitter posts shifted from one of discourse to one of public service announcements.

Notes

1 Metadata for each tweet included information such as number of tweets and followers for the account, time and location, information on posting device and other data provided through the API.
2 Number of users represents the number of individual accounts who used #safestampede or the phrase 'safe stampede' during the time in question.
3 Pseudonym
4 r/Calgary is the regional discussion board for Reddit.

References

Allen, M. (1998). *Rodeo cowboys in the North American imagination*. Reno, NV: University of Nevada Press.
Borra, E., & Rieder, B. (2014). Programmed method: Developing a toolset for capturing and analyzing tweets. *Aslib Journal of Information Management*, *66*(3), 262–278. doi:10.1108/AJIM-09-2013-0094
Bourgon, L. (2015). Herd mentality: Maisonneuve: A quarterly of arts, opinion & ideas. *67*, June 30 [Online]. Retrieved from https://maisonneuve.org/article/2015/06/11/herd-mentality/
Cammaerts, B. (2018). *The circulation of anti-austerity protest*. Cham: Palgrave Macmillan. doi:10.1007/978-3-319-70123-3
CBC. (2017). Calgary stampede 2017 attendance tops 1.2M. *CBC News*, July 17. Retrieved June 16, 2018, from www.cbc.ca/news/canada/calgary/calgary-stampede-2017-attendance-1.4209224
Foran, M. (Ed.). (2008). *Icon, brand, myth: The Calgary stampede*. Edmonton, AB: Athabasca University Press.
Kriesi, H. (1996). The organizational structure of new social movements in a political context. In D. McAdam et al. (Eds.), *Comparative perspectives on social movements* (pp. 152–184). Cambridge: Cambridge University Press.
Pavan, E. (2017). The integrative power of online collective action networks beyond protest: Exploring social media use in the process of institutionalization. *Social Movement Studies*, *16*(4), 433–446. doi:10.1080/14742837.2016.1268956
Spence, B. (2018). The WiseGuyz story. *Centre for Sexuality* [Online]. Retrieved September 12, 2018, from www.centreforsexuality.ca/programs-workshops/wiseguyz/
Tarrow, S. G. (2011 [1994]). *Power in movement: Social movements and contentious politics* (3rd ed.). Cambridge: Cambridge University Press.

Index

alcohol: consumption 40, 49, 70, 72–74, 81–82, 87–93, 97, 111; influence 49, 6, 12, 16, 79; *see also* intoxication
Argentina 5, 133, 135–137, 139–142
assault *see* sexual assault
assemblage 5, 70–73, 75, 79–82
Australia 74, 122, 135

Basque Country 4, 9–10, 12–18
blame 34, 49, 77; *see also* victim blaming
bullying 104–112
Butler, Judith 2, 134
bystander intervention 70, 80, 149, 155, 157–158, 161; bystander programmes 112

Calabar Carnival 40, 44, 47
Calgary Stampede (CS) 147–149, 153–155, 157–159; #safestampede 148–149, 151, 153–161
Canada 6, 17, 148
Carniriv festival 44, 47, 48
carnival 44, 49, 147; carnivalesque 72–73, 77, 81–82, 95, 97
circuit of protest 151, 154, 160
Coachella 69
colonialism 5, 58; anti-colonial 58; decolonial 57, 58–59, 66; postcolonial 58, 66
communitas 3, 7
cowboy 147, 157–159
Crenshaw, K. 2, 58
crime 5, 14, 51, 86–91, 93–98; categorisation of 10, 12, 24, 29–34, 36n8; reporting of 39–41
critical event studies 1–2, 4, 122
crowd: behaviours 24, 29, 40, 78–81, 138; management 127; spaces 35, 48–49, 69–70, 95, 133, 147

discrimination: of marginal groups 12, 73, 87, 93, 111; structural 5, 27, 124–125
drugs 16, 69, 72–73, 77, 111; *see also* intoxication

employees 104–106, 109–112
empowerment 7, 28, 65, 133, 139, 141
Equalities and Human Rights Commission 104–105, 110, 113
equality 1, 5, 10, 12–17, 28, 65, 119–129, 136, 141–142
ethnicity 2, 56, 87, 93, 108, 135
European Union 111

fear: culture of 18–20; perception of 27–28, 96; strategies to deal with 20, 34–35, 88–89, 157–160; women's fear 11–12, 16, 46, 125–126
feminism: decolonial 57–59; institutional 9–12, 16–21; rise of 21, 30–31
feminist: analysis 33–35, 40, 50–51, 106; movements 4–5, 20, 24–28, 30–31, 104, 133, 137–142, 151; scholarship 86–87, 91, 96, 107, 122, 134
festivals: arts 5, 29, 54–56, 59–61, 63–64, 66; commodification of 97, 120; cultural 5, 39–42, 44–45, 49–51; fiesta 5, 24–29, 33, 35, 44, 46–48; music 4–6, 40, 45–46, 69–74, 79–81, 86, 94–98, 104–113, 133–134, 137–141

gender ratio 109, 112
gender socialisation 33, 50
gig economy 109
Global North 135–136, 142
Global South 134–136, 142
grassroots 6, 63, 147–161
GRL PWR festival 140–142

Index

harassment *see* sexual harassment
heterosexism 29; heteropatriarchy 27
Hull 119–124, 127–130

inequality: festivals 1–5, 80; gender 20, 27, 35, 57, 73, 87, 91, 105–108, 134, 139–140; structural 5–6, 112, 119–130; *see also* equality
injustice: epistemic injustice 55–66; hermeneutical injustice 59; structural injustice 124–125
intersectionality 2, 56–59, 93, 134
International Women's Day 121
intoxication 77, 81, 90, 94–95
Intwasa International Arts Festival 54–66

Kaduna festival 40, 45

Lagos Fiesta 44–48
La Manada *see* Wolf Pack
leisure: deviant leisure 95–96; leisure spaces 89–98
LGBTQ+ 70, 73, 119
liminal/liminoid 1–4, 26, 72–83, 95–97, 124, 129; liminal-norm 2–3
Lorde, A. 7

male gaze 27, 93
masculinity: characteristics of 10–11, 20, 92–95; demonstrations of 25–28; dominance of 33–35; hegemonic 61–62, 73, 78–81; hypermasculinity 92–93, 148; space 133–134
#MeToo 21, 30, 104
misogyny 139, 159, 6

Nigeria 5, 39–52
night-time economy (NTE) 72, 90–94, 98

patriarchy: conceptions of 2, 7, 25; heteropatriarchy 27; logic of 5, 24, 27, 33–35; patriarchal society 10–12, 25, 27, 31, 50–57, 60–66, 87, 137–142
policy: analysis 94, 105; change 160; recommendations 51, 105, 111
power: gender relations 1–6, 20, 87, 134; heterosexual 93; male 25, 87, 142; multi-vectors of 58–59; societal structures 1, 10–13, 25–29, 40, 71–73, 104–112

race 58–60, 98
rape 2, 4, 9–11, 16, 39–51, 105, 111, 138; rape culture 5, 24–26, 30–35, 72

representation 11, 25–28, 63, 151, 154, 156–159; under-representation 55–57, 66; unequal representation 124
resistance 3, 5, 12, 28–34, 59, 95, 133–142
risk: behaviours 72–73, 93–94; festival spaces 6, 39, 49, 97, 119–120; gendered 77, 86–98; heteronormative 94; perceptions of 55, 76, 97; workplace 106, 110
rodeo 147–148, 160

San Fermin 9, 14, 24–35
safety 5, 46, 55, 63, 74, 86–89, 92–98, 157–158; safe spaces 62, 66, 97, 135, 137–139, 148
sexism: cultures of 60–61, 93; festivals 133, 139; resistance to 142, 148, 154; sexist behaviour/aggression/violence 12–21, 25–30; structural 5, 135; workplace 105
sexual assault: definitions of 24, 31–32, 36n8; at festivals 2, 30, 46, 49, 69, 96, 137; justice system 26, 33–34; motivations 40; responses to 10, 31, 36n5, 75
sexual harassment: definitions of 13, 42; fear of 60; at festivals 44–49, 64, 66, 69, 76–77, 80, 104–112; justice system 41; normalisation of 6, 155–160; perceptions of 54; in public spaces 87–88, 91, 93, 96, 133; resistance to 148–149; societal shaping of 40–41
silence (culture of) 40–42, 50–51, 64
social media: Tumblr 149, 154–157; Twitter 148–161; YouTube 158
social movement organisations (SMO) 150
space: festival spaces 1–6, 28–35, 70–73, 76–77, 82, 94–98, 124; public space 2–5, 11, 16–20, 24–35, 62, 71, 77, 86–89, 93, 134; social space 34, 63, 86; spatial arrangement 72, 77, 127–129
Spain 4, 9–21, 24–36
Spivak, G.C. 55, 58, 66; *see also* subaltern
stereotypes 16, 24, 27, 39, 41, 48–51, 57, 159
structural disputes 119–130
subaltern 55, 66

transgression 3, 69, 72, 90, 94–97
transphobia 29
Turner, V. 2–3

UK City of Culture 119
United Nations Committee on the Discrimination Against Women 111

victim blaming 2, 33, 39, 134
victimisation 40–51, 80, 88–89, 107, 111
violence: epistemic violence 5, 55, 63, 66; gender-based violence (GBV) 104–111; gendered-violence 1–7, 13–17, 51, 54–56, 60, 64–66, 70, 139; sexual violence 5, 9–14, 24–35, 39–51, 64, 69–82, 86, 88, 96–98, 105; structural violence 120, 124, 126, 129; verbal violence 9

Wolf Pack 4, 10, 16, 21, 24–30
Women and Equalities Committee 104–105
Women of the World Festival (WOW) 119–129
women's bodies 2, 10, 12, 21, 25, 33–34, 81, 88
World Health Organisation 105, 111

Zimbabwe 5, 54–66

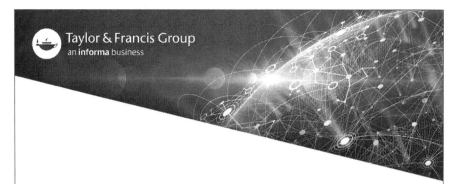